AS/A-LEVEL YEAR 1
STUDENT GUIDE

EDEXCEL

Chemistry

Topics 6–10

Organic chemistry I
Modern analytical techniques I
Energetics I
Kinetics I
Equilibrium I

George Facer

Rod Beavon

PHILIP ALLAN FOR
HODDER
EDUCATION
AN HACHETTE UK COMPANY

Philip Allan, an imprint of Hodder Education, an Hachette UK company, Blenheim Court, George Street, Banbury, Oxfordshire OX16 5BH

Orders

Bookpoint Ltd, 130 Milton Park, Abingdon, Oxfordshire OX14 4SB

tel: 01235 827827

fax: 01235 400401

e-mail: education@bookpoint.co.uk

Lines are open 9.00 a.m.–5.00 p.m., Monday to Saturday, with a 24-hour message answering service. You can also order through the Hodder Education website: www.hoddereducation.co.uk

© George Facer and Rod Beavon 2015

ISBN 978-1-4718-4354-9

First printed 2015

Impression number 5 4 3 2 1

Year 2019 2018 2017 2016 2015

This guide has been written specifically to support students preparing for the Edexcel AS and A-level Chemistry examinations. The content has been neither approved nor endorsed by Edexcel and remains the sole responsibility of the authors.

Cover photo: TTstudio/Fotolia

Typeset by Integra Software Services Pvt. Ltd, Pondicherry, India

Printed in Italy

Hachette UK's policy is to use papers that are natural, renewable and recyclable products and made from wood grown in sustainable forests. The logging and manufacturing processes are expected to conform to the environmental regulations of the country of origin.

Contents

Content Guidance

Questions & Answers

■Getting the most from this book

Exam tips

Advice on key points in the text to help you learn and recall content, avoid pitfalls, and polish your exam technique in order to boost your grade.

Knowledge check

Rapid-fire questions throughout the Content Guidance section to check your understanding.

Knowledge check answers

1 Turn to the back of the book for the Knowledge check answers.

Summaries

■ Each core topic is rounded off by a bullet-list summary for quick-check reference of what you need to know.

Exam-style questions

Commentary on the questions

Tips on what you need to do to gain full marks, indicated by the icon (e)

Sample student answers

Practise the questions, then look at the student answers that follow.

Commentary on sample student answers

Find out how many marks each answer would be awarded in the exam and then read the comments (preceded by the icon (e)) showing exactly how and where marks are gained or lost.

Questions & Answers

Question 6

The reaction between sulfur dioxide and oxygen in a closed system is in dynamic equilibrium:

$2SO_2(g) + O_2(g) \rightleftharpoons 2SO_3(g)$ $\qquad \Delta_r H = -196\,kJ\,mol^{-1}$

(a) (i) Explain what is meant by *dynamic equilibrium*. [2 marks]

(e) Remember to define both *dynamic* and *equilibrium*.

(ii) The correct expression for K_c for this reaction is: [1 mark]

A $\dfrac{[SO_3]}{[SO_2][O_2]}$ B $\dfrac{[SO_3]^2}{[SO_2]^2[O_2]}$ C $\dfrac{[SO_2]^2[O_2]}{[SO_3]}$ D $\dfrac{[SO_3]^2[O_2]}{[SO_2]^2}$

(b) State the effect on the position of equilibrium of this reaction of:
(i) increasing the temperature [1 mark]
(ii) increasing the pressure [1 mark]

(e) There is no need for any explanation.

(c) This reaction is the first step in the industrial production of sulfuric acid. A temperature of 450°C, a pressure of 2 atm and a catalyst are used. Justify the use of these conditions:
(i) a temperature of 450°C [2 marks]
(ii) a pressure of 2 atm [2 marks]
(iii) a catalyst [1 mark]

(e) For each, a comment on the *rate* and another on the *yield* is needed.

(d) Name the catalyst used industrially. [1 mark]

(e) Explain why an industrial manufacturing process such as this cannot be in equilibrium. [2 marks]

[Total: 13 marks]

Student answer

(a) (i) The forward and backward reactions are occurring at the same rate ✓, so there is no net change in composition ✓.

(e) The first mark is for *dynamic* and the second for *equilibrium*.

(ii) B

(e) A is for the equation $SO_2 + \frac{1}{2}O_2 \rightleftharpoons SO_3$. C and D are for the reverse reactions.

(b) (i) Moves to the left or concentration of products decreases ✓
(ii) Moves to the right or concentration of products increases ✓

86 Edexcel Chemistry

■ About this book

This guide is the second of a series covering the Edexcel specification for AS and A-level chemistry. It offers advice for the effective study of topics 6–10, which are examined on AS paper 2 (together with topics 2 and 5 which are covered in the student guide covering topics 1–5 in this series) and on A-level papers 1, 2 and 3. The aim of this guide is to help you understand the chemistry — it is not intended as a shopping list, enabling you to cram for an examination. The guide has two sections:

- The **Content Guidance** is not intended to be a textbook. It offers guidelines on the main features of the content of topics 6–10, together with advice on making study more productive.
- The **Questions & Answers** section starts with an introduction that gives advice on approaches and techniques to ensure you answer the examination questions in the best way you can. It then provides exam-style questions with student answers and comments on how the answers would be marked.

The effective understanding of chemistry requires time. No one suggests that chemistry is an easy subject, but if you find it difficult you can overcome your problems by the proper investment of your time.

To understand the chemistry, you have to make links between the various topics. The subject is coherent and is not a collection of discrete modules. Once you have spent time thinking about chemistry, working with it and solving chemical problems, you will become aware of these links. Spending time this way will make you fluent with the ideas. Once you have that fluency, and practise the good techniques described in this book, the examination will look after itself. Don't be an examination automaton — be a chemist.

The specification

The specification describes the chemistry that can be tested in the examinations and the format of those exams. It can be obtained from Edexcel, either as a printed document or from the internet at www.edexcel.com.

Learning to learn

Learning is not instinctive — you have to develop suitable techniques to make effective use of your time. In particular, chemistry has peculiar difficulties that need to be understood if your studies are to be effective from the start.

Planning

Efficient people do not achieve what they do by approaching life haphazardly. They plan — so that if they are working, they mean to be working, and if they are watching television, they have planned to do so. Planning is essential. You must know what you have to do each day and set aside time to do it. Furthermore, to devote time to study means you may have to give something up that you are already doing. There is no way that you can generate extra hours in the day.

Be realistic in your planning. You cannot work all the time and you must build in time for recreation and family responsibilities.

Targets

When devising your plan, have a target for each study period. This might be a particular section of the specification, or it might be rearranging of information from text into pictures, or drawing a flowchart relating all the reactions of alkanes and alkenes. Whatever it is, be determined to master your target material before you leave it.

Reading chemistry textbooks

A page of chemistry may contain a range of material that differs widely in difficulty. Therefore, the speed at which the various parts of a page can be read may have to vary, if it is to be understood. You should read with pencil and paper to hand and jot things down as you go — for example, equations, diagrams and questions to be followed up. If you do not write down the questions, you will forget them; if you do not master detail, you will never become fluent in chemistry.

Practising skills

Chemical equations

Equations are used because they are quantitative, concise and internationally understood. Take time over them, copy them and check that they balance. Most of all, try to visualise what is happening as the reaction proceeds. If you can't, make a note to ask someone who can or — even better — ask your teacher to *show* you the reaction if possible. Chemical equations describe real processes; they are not abstract algebraic constructs.

Graphs

Graphs give a lot of information, and they must be understood in detail rather than as a general impression. Take time over them. Note what the axes are, what the units are, the shape of the graph and what the shape means in chemical terms.

Tables

These are a means of displaying a lot of information. You need to be aware of the table headings and the units of numerical entries. Take time over them. What trends can be seen? How do these relate to chemical properties? Sometimes it can be useful to convert tables of data into graphs.

Diagrams

Diagrams of apparatus should be drawn in section. When you see them, copy them and ask yourself why the apparatus has the features it has. What is the difference between a distillation and a reflux apparatus, for example? When you do practical work, examine each piece of apparatus closely so that you know both its form and its function.

Calculations

Do not take calculations on trust — work through them. First, make certain that you understand the problem and then that you understand each step in the solution. Make clear the units of the physical quantities used and make sure you understand the underlying chemistry. If you have problems, ask.

Always make a note of problems and questions that you need to ask your teacher. Learning is not a contest or a trial. Nobody has ever learnt anything without effort or without running into difficulties from time to time — not even your teachers.

Notes

Most people have notes of some sort. Notes can take many forms: they might be permanent or temporary; they might be lists, diagrams or flowcharts. You have to develop your own styles. For example, notes that are largely words can often be recast into charts or pictures and this is useful for imprinting the material. The more you rework the material, the clearer it will become.

Whatever form your notes take, they must be organised. Notes that are not indexed or filed properly are useless, as are notes written at enormous length and those written so cryptically that they are unintelligible a month later.

Writing

In chemistry, extended writing is often not required. However, you need to be able to write concisely and accurately. This requires you to marshal your thoughts properly and needs to be practised during your day-to-day learning.

Have your ideas assembled in your head before you start to write. You might imagine them as a list of bullet points. Before you write, have an idea of how you are going to link these points together and also how your answer will end. The space available for an answer is a poor guide to the amount that you have to write — handwriting sizes differ hugely, as does the ability to write succinctly. Filling the space does not necessarily mean you have answered the question. The mark allocation suggests the number of points to be made, not the amount of writing needed.

Content Guidance

∎ Topic 6 Organic chemistry I

Topic 6A Introduction to organic chemistry

Organic chemistry is dominated by a number of **homologous series**, arising from the presence of one or more functional groups (see below) in a molecule. The huge variety of organic compounds arises from the fact that organic chemistry is rather like atomic Lego, as you will realise if you have used molecular model kits.

Exam tip

Most organic compounds burn in excess air to form carbon dioxide and water.

Homologous series

A homologous series is a series of compounds that:

- have a common general formula
- differ by CH_2
- show a trend in physical properties, for example boiling temperature
- show similar chemical properties since all members of the same homologous series have the same functional group

The **alkenes** are an example of a homologous series in that:

- their general formula is C_nH_{2n} (though not all compounds with this general formula are alkenes)
- they have a C=C double bond (which other compounds with the same general formula do not have)
- there is no alkene with only one carbon atom since there could not be a C=C bond

The first three straight-chain compounds in the series are ethene, $CH_2{=}CH_2$, propene, $CH_3CH{=}CH_2$, and butene. Butene has structural and geometric isomers (see pp. 13–15), one of which is but-1-ene, $CH_2{=}CHCH_2CH_3$. The boiling temperatures of these three alkenes are $-104°C$, $-47.7°C$ and $-6.2°C$ respectively. All three react rapidly with bromine water, with potassium manganate(VII) solution, and with all the other reagents given on pages 22–23.

Functional groups

A **functional group** is a small group of atoms or perhaps a single atom that determines the chemistry of a molecule. The homologous series and functional groups required for AS and A-level are given in Table 1. The atoms in bold are the functional groups and R represents an organic group.

Series	General formula	First member
Alkanes	C_nH_{2n+2}	CH_4, methane
Alkenes	C_nH_{2n}	$CH_2{=}CH_2$, ethene
Halogenoalkanes	R**X** where **X** is –Cl, –Br or –I	CH_3Cl, chloromethane (or bromo- or iodo-)
Primary alcohols	R**CH₂OH**	CH_3OH, methanol

→

Series	General formula	First member
Secondary alcohols	RR′**CHOH** OH $\text{H}-\overset{\text{OH}}{\underset{\text{R}'}{\text{C}}}\cdots\text{R}$ R and R′ may be the same	$CH_3CH(OH)CH_3$, propan-2-ol
Tertiary alcohols	RR′R″**COH** OH $\text{R}''-\overset{\text{OH}}{\underset{\text{R}'}{\text{C}}}\cdots\text{R}$ The Rs may be the same	$(CH_3)_3COH$, 2-methylpropan-2-ol
Aldehydes	R**CHO** R $\underset{\text{H}}{\text{C}}=\text{O}$	HCHO, methanal
Ketones	R**COR′** R $\underset{\text{R}'}{\text{C}}=\text{O}$ R and R′ can be the same	CH_3COCH_3, propanone
Carboxylic acids	R**COOH** R $\underset{\text{HO}}{\text{C}}=\text{O}$	HCOOH, methanoic acid
Acid chlorides	R**COCl** R $\underset{\text{Cl}}{\text{C}}=\text{O}$	CH_3COCl, ethanoyl chloride
Esters	R**COOR′** R $\underset{\text{R}'\text{O}}{\text{C}}=\text{O}$ R and R′ can be the same. R, but not R′, can be H	$HCOOCH_3$, methyl methanoate
Primary amines	R**NH$_2$**	CH_3NH_2, aminomethane

Table 1 Homologous series and functional groups for AS and A-level

Naming and drawing organic compounds

In organic chemistry, the structure of a compound is central to its chemistry. All but rather simple molecular formulae give rise to a variety, sometimes enormous, of different structures.

Nomenclature

The International Union of Pure and Applied Chemistry (IUPAC) has generated a series of rules giving organic compounds **systematic names**. Knowing the rules and the systematic name of a compound means the formula can be written, and vice versa.

The names may be long if the compounds are large. Systematic names are used in the Edexcel specification for most organic substances, though there are exceptions, for example the amino acid, glycine. However systematic names are less common in chemistry beyond school level. This is partly because American textbooks and chemical suppliers do not often use them and partly because the names are unwieldy if more than five or six carbon atoms are involved.

The rules for naming alkanes are given below, with some examples. These rules are built on to name other types of organic molecule.

- Identify the longest carbon chain — take care with this since structures can be written with 90° bond angles that may mask the longest chain. Thus the compound shown in Figure 1 has a four-carbon longest chain, not three carbons.
- The name is based on the alkane with the same number of carbon atoms as the longest chain.
- Substituent groups have the name ending changed to -yl.

This is illustrated for alkanes up to C_5 in Table 2.

C	Alkane	Name stem	Substituent group	Structure
1	Methane, CH_4	Meth-	Methyl	CH_3-
2	Ethane, C_2H_6	Eth-	Ethyl	CH_3CH_2-
3	Propane, C_3H_8	Prop-	Propyl	$CH_3CH_2CH_2-$
4	Butane, C_4H_{10}	But-	Butyl	$CH_3CH_2CH_2CH_2-$
5	Pentane, C_5H_{12}	Pent-	Pentyl	$CH_3CH_2CH_2CH_2CH_2-$

Table 2 Naming alkanes

- The position of each substituent group is indicated by a number. This corresponds to the position in the chain of the carbon atom to which it is attached, so that the number of the substituent group is the lowest possible. However, for substances such as carboxylic acids, RCOOH, where the functional group is at the end of a chain, the carbon atoms are always numbered from that end.
- Different series of compounds have the name-ending modified to indicate which homologous series is involved.
- Where two groups that are the same are substituted on a given carbon atom, the number is repeated and the substituent is prefixed 'di'.
- Where two groups that are the same are on different carbon atoms, 'di' is still used. If there are three groups the same on different carbons then 'tri' is used, if there are four groups then 'tetra' is used, and so on.

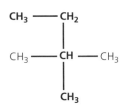

Figure 1

Knowledge check 3

Name $CH_3CH=C(CH_3)CH(CH_3)CH_2CH_3$.

Examples of systematic names

Alkanes

See Figure 2.

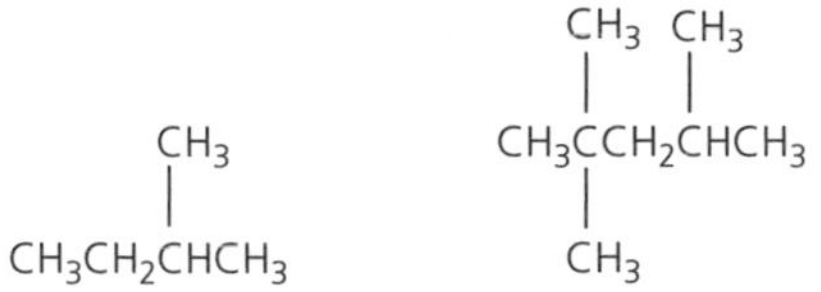

$$CH_3CH_2CHCH_3$$
$$\overset{\displaystyle CH_3}{|}$$
2-methylbutane

$$CH_3CCH_2CHCH_3$$
2,2,4-trimethylpentane

Figure 2

Alkenes

The position of the double bond (or bonds) is given by the number of the carbon atom that the double bond starts from (Figure 3).

$CH_3CH=CH_2$
Propene

$CH_3CH_2CH=CH_2$
But-1-ene

$CH_3CH=CHCH_3$
But-2-ene

$CH_2=CH–CH=CH_3$
Buta-1,3-diene;
not an alkene, but an alkadiene

Figure 3

Halogenoalkanes

See Figure 4.

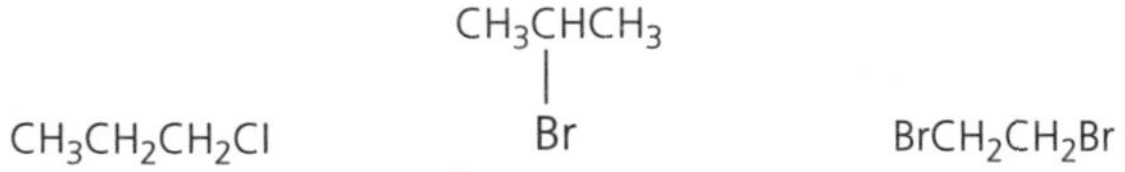

$CH_3CH_2CH_2Cl$
1-chloropropane

CH_3CHCH_3
$|$
Br
2-bromopropane

$BrCH_2CH_2Br$
1,2-dibromoethane

Figure 4

Alcohols

See Figure 5.

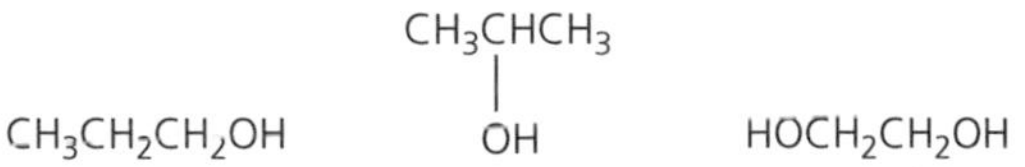

$CH_3CH_2CH_2OH$
Propan-1-ol

CH_3CHCH_3
$|$
OH
Propan-2-ol

$HOCH_2CH_2OH$
Ethane-1,2-diol

Figure 5

Aldehydes

See Figure 6.

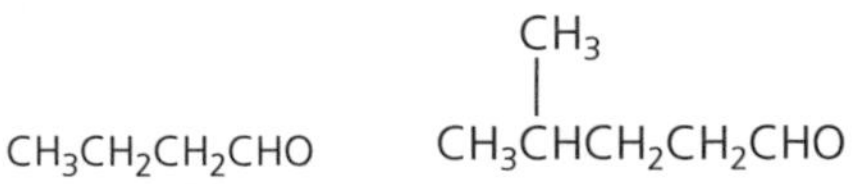

$CH_3CH_2CH_2CHO$
Butanal

$CH_3CHCH_2CH_2CHO$
$\overset{\displaystyle CH_3}{|}$
4-methylpentanal

Figure 6

Ketones

See Figure 7.

CH_3COCH_3 $CH_3COCH_2CH_2CH_3$ $CH_3CH_2COCH_2CH_3$
Propanone   Pentan-2-one   Pentan-3-one

Figure 7

Carboxylic acids

See Figure 8.

$CH_3CH_2CH_2COOH$ $CH_2 = CHCOOH$
  Butanoic acid   Propenoic acid

$CH_3CH = CH_2COOH$ CH_3CH_2COOH
  But-2-enoic acid   Propanoic acid — note the difference

CH_3CHCH_2COOH HOOC–COOH
       | Ethanedioic acid
       OH
3-hydroxybutanoic acid

Figure 8

Drawing organic molecules

- The **empirical formula** of an organic molecule shows the simplest ratio of the atoms of its component elements.
- The **molecular formula** shows the actual number of atoms of each element.

Butane has the empirical formula C_2H_5 and the molecular formula C_4H_{10}. The molecular formula of an organic compound may not be particularly useful, since there is usually more than one possible structure — for example, C_4H_{10} has three possible structures. This problem is overcome by presenting formulae in different ways:

- **Structural formulae** show most of the structure, but not all of the bonds are shown. The side-chain substituents are shown in parentheses.
- **Displayed formulae** show all of the bonds.
- **Skeletal formulae** are used widely for natural products and other large molecules. They show the carbon skeleton by lines (which represent the bonds joining the atoms). Hydrogen atoms, unless bonded to oxygen or nitrogen, are not shown. All other types of atom are shown.

Consider the molecule 2-methylbutane. Four representations of its structure are shown in Figure 9.

> **Exam tip**
>
> The structural formula with the side-chain substituents in parentheses is useful for typing on a single line. The semi-displayed formula would be used in hand-written material.

> **Exam tip**
>
> Never use an ambiguous molecular formula such as $C_2H_4Br_2$—use a structural formula, such as CH_3CHBr_2 or CH_2BrCH_2Br.

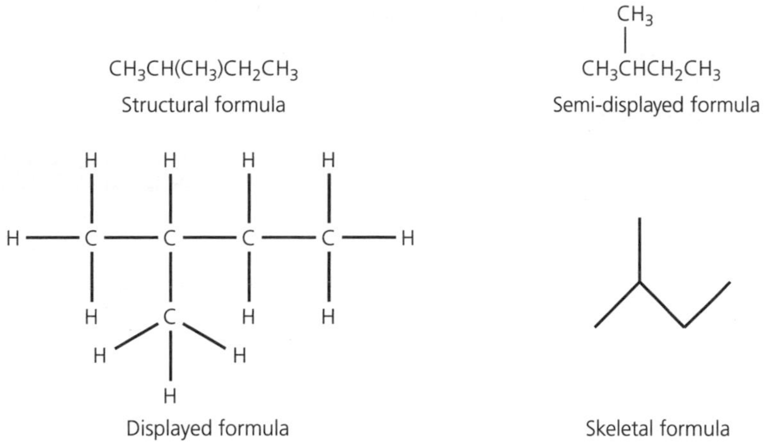

CH$_3$CH(CH$_3$)CH$_2$CH$_3$
Structural formula

CH$_3$
|
CH$_3$CHCH$_2$CH$_3$
Semi-displayed formula

Displayed formula

Skeletal formula

Figure 9 Representations of 2-methylbutane structure

The example of *cis*-but-2-enoic acid illustrates how to represent C=C double bonds and atoms other than carbon (Figure 10).

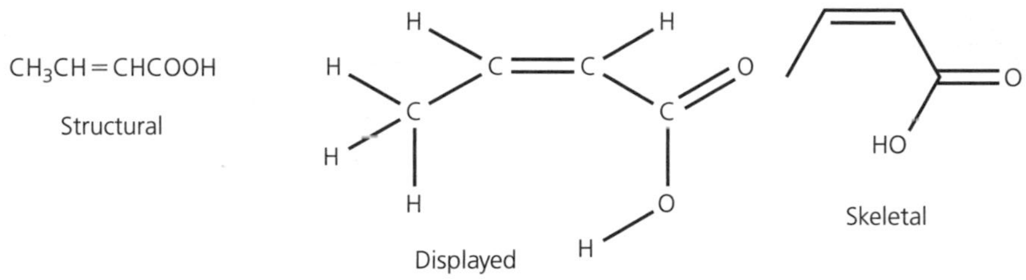

CH$_3$CH=CHCOOH

Structural

Displayed

Skeletal

Figure 10 *cis*-but-2-enoic acid

Isomerism

There are two different types of isomerism required at AS.

Structural isomerism

Structural isomers are compounds that have the same molecular formula but different structural formulae. The first three alkanes, CH$_4$, CH$_3$CH$_3$ and CH$_3$CH$_2$CH$_3$, have only one possible structure in each case. However the fourth, CH$_3$CH$_2$CH$_2$CH$_3$, has a structural isomer in which the atoms are arranged differently (Figure 11).

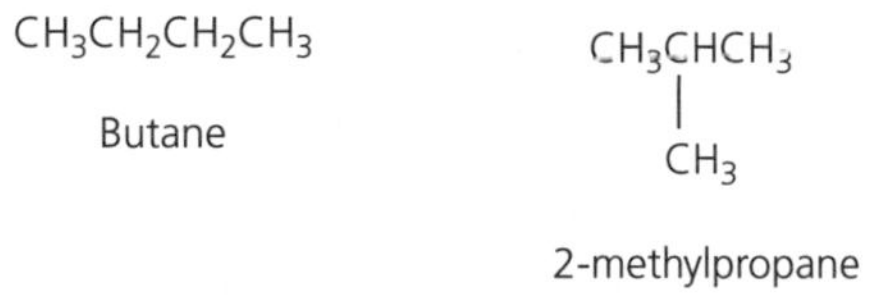

CH$_3$CH$_2$CH$_2$CH$_3$

Butane

CH$_3$CHCH$_3$
|
CH$_3$

2-methylpropane

Figure 11 Butane and 2-methylpropane are structural isomers

In the case of the alkanes, the structural isomers are also alkanes. However, structural isomers may be of different types. The molecular formula C$_2$H$_6$O could represent ethanol, CH$_3$CH$_2$OH, or methoxymethane, CH$_3$OCH$_3$, which is an ether. This is sometimes called **functional group isomerism**.

The number of structural isomers rises rapidly with the number of carbon atoms in the compound. For example, pentane has two other structural isomers, so there are three structures that have the molecular formula C_5H_{12} (Figure 12). The numbers of isomers of some other alkanes are shown in Table 3.

$CH_3CH_2CH_2CH_2CH_3$

$CH_3CHCH_2CH_3$
|
CH_3

$\begin{array}{c} CH_3 \\ | \\ CH_3CCH_3 \\ | \\ CH_3 \end{array}$

Pentane   2-methylbutane   2,2-dimethylpropane

Figure 12 Structural isomers of pentane

Knowledge check 4

Draw the skeletal formulae of the three isomers of molecular formula C_5H_{12}.

Alkane	Number of isomers
C_8H_{18}	18
$C_{10}H_{22}$	75
$C_{12}H_{26}$	355
$C_{14}H_{30}$	1858
$C_{20}H_{42}$	366319
$C_{25}H_{52}$	36797588
$C_{30}H_{62}$	4111846763
$C_{40}H_{82}$	62491178805831

Table 3 Numbers of isomers of the alkanes

Exam tip

Because the number of potential structures can be large (even for small compounds), you should always use the structural formula to eliminate ambiguity.

Geometric isomerism

Geometric isomerism is a subset of stereoisomerism. The isomers have different orientation in space and it can result from restricted rotation about a carbon–carbon double bond, provided that the groups on a given carbon in the C=C bond are not the same. Thus in Figure 13 'a' and 'e' must be different, as must 'b' and 'd'. The groups do not *all* have to be different from one another.

Figure 13

Figure 14 shows two geometric isomers. These are *cis-* and *trans-*1,2-dichloroethene. In *cis*-isomers the substituent groups, in this case the chlorine groups, are on the same side of the C=C bond. In *trans* isomers the substituent groups are on opposite sides of the C=C bond.

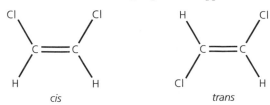

cis   *trans*

Figure 14 Isomers of 1,2-dichloroethene

Exam tip

Anything that prevents rotation, such as a ring of carbon atoms, could cause geometric isomerism. Thus 1,2-dichlorocyclohexane exists as two geometric isomers.

Sideways overlap of the p-orbitals to give a π-bond means that there is no rotation about C=C bonds, except at high temperatures (Figure 15).

E–Z system of naming geometric isomers

If there are only two substituents (other than hydrogen) on the C=C bond, the *cis–trans* naming system works well. In the case of compounds that have more than two different groups around the C=C bond — for example, 2-bromobut-2-ene, the two forms of which are shown in Figure 16 — it is better to name them using a different system — the *E–Z* system.

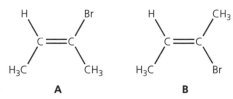

Figure 16 Isomers of 2-bromobut-2-ene

In the *E–Z* system the substituent groups are ordered in a priority determined by the atomic mass of the atom directly bonded to a carbon atom. Taking isomer A above, on the left-hand carbon, the carbon (atomic mass 12.0) in the methyl group has higher priority than the hydrogen (atomic mass 1.0). On the right-hand carbon, bromine (atomic mass 79.9) has higher priority than carbon (atomic mass 12.0). In isomer A the two high-priority substituents are, therefore, the methyl (CH_3) group and the bromine, which are on opposite sides of the C=C bond. This isomer is the *entgegen* isomer, or *E*-isomer.

In isomer B, the high-priority groups are on the same side of the C=C bond, and this is called the *zusammen* or Z-isomer.

In fact, 2-bromobut-2-ene could be named using the *cis–trans* nomenclature; isomer A, with the methyl groups on the same side, being the *cis*-isomer. If one of these methyl groups were changed to an ethyl group then it would not be possible to assign *cis*- or *trans*- to the isomers. If all four groups are different, as in F and G in Figure 17, then only the *E–Z* system will work. The substituent with the highest priority on the left-hand carbon is iodine and that on the right-hand carbon is chlorine. These are on the same side of the double bond and so F is the Z-isomer.

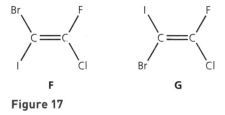

F   **G**

Figure 17

If the groups get larger (e.g. ethyl or propyl), the rules are extended to the masses of the groups attached to the double bonded carbon atoms.

Figure 15

Knowledge check 5

Explain why but-2-ene, $CH_2=CHCH_3$, does not show geometric isomerism.

Knowledge check 6

Draw the skeletal formula of the *E*-isomer of 3-methylpent-2-ene.

Types of reaction

Addition reactions

Addition reactions occur when two or more molecules combine to give one product:

$$H_2C=CH_2 + H_2 \rightarrow CH_3CH_3$$

$$H_2C=CH_2 + HBr \rightarrow CH_3CH_2Br$$

$$n(H_2C=CH_2) \rightarrow (-CH_2CH_2-)_n$$

The last example is of **addition polymerisation**.

Elimination reactions

Elimination reactions occur when a molecule loses two or more atoms, turning a single into a double bond. For example, adding KOH to hot bromoethane in ethanol eliminates hydrogen bromide:

$$CH_3CH_2Br + KOH \rightarrow H_2C=CH_2 + KBr + H_2O$$

Substitution reactions

Substitution reactions involve the replacement of an atom or group of atoms with another atom or group. Halogenoalkanes undergo **nucleophilic substitution**:

$$CH_3CH_2Br + KOH \rightarrow CH_3CH_2OH + KBr$$

$$CH_3CH_2Br + KCN \rightarrow CH_3CH_2CN + KBr$$

Oxidation

Oxidation occurs when a species loses electrons. In organic chemistry at A level, oxidation is usually interpreted as a gain of oxygen or the loss of hydrogen:

$$CH_3CH_2OH \rightarrow CH_3CHO + H_2 \qquad \text{(over hot Cu catalyst)}$$

Oxidising agents such as potassium dichromate(VI) are usually represented in equations as [O], as in the reaction of acidified potassium dichromate(VI) with ethanol:

$$CH_3CH_2OH + [O] \rightarrow CH_3CHO + H_2O$$

Reduction

Reduction occurs when a species gains electrons. In organic chemistry at A level, reduction is usually interpreted as a gain of hydrogen or the loss of oxygen. Reducing agents other than hydrogen itself are represented as [H].

$$H_2C=CH_2 + H_2 \rightarrow CH_3CH_3$$

$$CH_3CHO + 2[H] \rightarrow CH_3CH_2OH$$

Hydrolysis

Hydrolysis is nominally a reaction with water, though most hydrolysis reactions are catalysed by acid or alkali (or by enzymes). The hydrolysis of esters is typical. Acid hydrolysis gives an equilibrium mixture.

Exam tip
You must be able to classify reactions, especially addition, elimination and substitution reactions.

Exam tip
When [O] or [H] are used they must be preceded by the correct number.

$$CH_3COOCH_2CH_3 + H_2O \rightleftharpoons CH_3COOH + CH_3CH_2OH \text{ (acid hydrolysis)}$$

but alkaline hydrolysis is not reversible

$$CH_3COOCH_2CH_3 + NaOH \rightarrow CH_3COONa + CH_3CH_2OH \text{ (alkaline hydrolysis)}$$

Polymerisation

Polymerisation is a reaction in which a large number of monomer molecules are joined to make a long-chain molecule with a molar mass in the hundreds of thousands. Alkenes undergo **addition polymerisation**:

$$n(H_2C{=}CH_2) \rightarrow (-CH_2CH_2-)_n$$

Topic 6B Alkanes

- **Alkanes** (also known as the paraffins) are hydrocarbons — compounds that contain hydrogen and carbon *only*.
- Their molecules are saturated — they contain only single bonds.
- They form a homologous series, with the general formula C_nH_{2n+2}.

The first four alkanes in the series are:

- methane, CH_4
- ethane, CH_3CH_3
- propane, $CH_3CH_2CH_3$
- butane, $CH_3CH_2CH_2CH_3$

For alkanes with five or more carbon atoms, the names are related to the number of carbon atoms in the chain: pentane, hexane, heptane, octane etc.

Cycloalkanes are compounds containing a ring (often of 6) carbon atoms. Their general formula is C_nH_{2n} and they react in a similar way to alkanes. Examples are cyclohexane, C_6H_{12} and cyclobutane, C_4H_8.

Fuels

Fuels are obtained from the **fractional distillation** of crude oil. This process usually produces small amounts of the petrol and diesel fractions and larger amounts of the heavier fuel and lubricating oil fractions. The latter are passed over suitable catalysts, with heating, in order to **crack** (or break) the molecules into smaller alkanes and alkenes. Alkanes can also be **reformed** by passing them over a suitable catalyst. Straight-chain molecules are converted to branched alkanes and cycloalkanes which have a much higher octane rating and so are more suitable for petrol-driven cars. The alkenes (see page 21) produced in cracking are used for chemical synthesis, as they are too valuable to burn.

Pollution problems

Combustion of hydrocarbon fuels produces a variety of pollutants. There is also some incomplete combustion, which produces carbon monoxide and particulates. The high temperatures in a car engine cause the oxygen and nitrogen in the air to combine to form nitrogen oxides, such as NO and NO_2. Heavy fuel oils, as used in marine engines, contain a significant proportion of sulfur compounds and, when burnt, produce sulfur dioxide. All these gases are harmful to the environment.

Petrol cars are fitted with catalytic converters. These convert the toxic carbon monoxide and nitrogen oxides into carbon dioxide and nitrogen.

$$CO + NO \rightarrow CO_2 + N_2$$

Liquid and gaseous fuels

The advantages and disadvantages of liquid and gaseous fuels depend on:

- whether the fuel has to be carried with the device using it (as in motor vehicles) or whether it can be piped (as in domestic gas supplies)
- any safety aspects specific to the particular fuel, especially the handling of potentially explosive gases (these do not include flammability, which some students seem to think is a problem — there are no non-flammable fuels)
- the energy yield per mole, volume or mass (shown in Table 4) (the molar volume of the gases at room temperature is taken as $24\,dm^3$).

Fuel	M_r	$\Delta_cH/kJ\,mol^{-1}$	$\Delta_cH/kJ\,cm^{-3}$	$\Delta_cH/kJ\,g^{-1}$
Hydrogen(g)	2	−286	−0.012	−143
Methane(g)	16	−890	−0.037	−55.6
Butane(g)	58	−2877	−0.12	−49.6
Butane(l)	58	−2877	−29.8	−49.6
Octane(l)	114	−5470	−33.8	−48.0
Ethanol(l)	46	−1367	−23.4	−29.7

Table 4 Energy yields of some fuels

Hydrogen and methane

Hydrogen and methane have a high energy yield per gram, but are not dense and therefore large volumes of gas are needed. These fuels are piped to the point of use.

Hydrogen can be used in cars and buses — it is compressed and absorbed in a suitable metal sponge. However, the range of such vehicles is limited and the relatively heavy gas canisters take up a lot of space. There are currently few hydrogen filling stations.

All gaseous fuels present special handling requirements to avoid leakage of flammable or explosive gases.

Environmental factors

Hydrogen burns to give water only. Remember, though, that the means of producing hydrogen (electrolysis of brine or reaction of methane with steam) consumes energy. The environmental advantages of burning hydrogen have to be balanced by the various environmental costs of its production. Methane is used in fixed systems and is burned efficiently, so that little carbon monoxide pollution results.

Butane

Butane has a high energy yield per gram, but the gas is not dense. It is easily compressed to give liquefied petroleum gas (LPG). Butane can be used in cars that are adapted for it — the number of filling stations selling it is rising. It is also used in tanks and cylinders as a constituent of Calor gas. The liquid form of butane vaporises readily and requires special handling facilities.

Exam tip

Do not state that hydrogen as a fuel has a zero carbon footprint. Huge amounts of CO_2 are produced in its manufacture.

Octane

Octane, a constituent of petrol, has a high energy yield per gram and is the densest of the fuels listed. Therefore, small volumes of the fuel can be carried in motor vehicles. It is universally available. The liquid is volatile, but not so volatile that special handling is needed.

Environmental factors

The yield of carbon monoxide from internal combustion engines, which use octane, is high. Most of this can be removed by the use of catalytic converters, in which carbon monoxide reacts with the nitrogen oxides that are also produced, to give nitrogen and carbon dioxide.

Ethanol

The alcohol ethanol has a high energy yield per gram, but it is not as dense as octane. Therefore, larger volumes of liquid are needed compared with octane. Ethanol is made from oil, if that is available, so it is not a cost-effective fuel for countries that have relatively cheap oil. Countries, such as Brazil, that have no oil but have a lot of sugar cane can make cheap, fuel-grade ethanol by fermentation. Car fuel in Brazil contains a high percentage of ethanol.

Environmental factors

Ethanol is a clean fuel, producing little carbon monoxide.

> **Exam tip**
>
> The production of bioethanol from starch (maize) has a high carbon footprint and reduces the world's food supply.

Reactions of alkanes

Reaction with oxygen

Alkanes burn in a plentiful supply of oxygen or air to give carbon dioxide and water. The reactions are strongly exothermic, hence the use of alkanes as fuels.

- Natural gas consists primarily of methane. It burns to give:

$$CH_4 + 2O_2 \rightarrow CO_2 + 2H_2O + heat$$

- Petrol is a mixture that can be represented by octane. It burns to give:

$$C_8H_{18} + 12\tfrac{1}{2}O_2 \rightarrow 8CO_2 + 9H_2O + heat$$

The main constituent in petrol is actually 2,2,4-trimethylpentane, which is an isomer of octane. In internal combustion engines, *complete* combustion does not occur; carbon and carbon monoxide are produced as well as carbon dioxide. In limited oxygen, all of the hydrogen in a given molecule is always oxidised — hydrogen is *never* a product. Carbon, seen as smoke, is also produced in limited oxygen.

Reaction with chlorine or bromine

In the presence of ultraviolet (UV) light or sunlight, alkanes react with chlorine or bromine to give halogenoalkanes. In strong UV or focused sunlight the reaction is explosive.

The reaction involves **radical substitution**. This produces a mixture of products, because radicals (in this case halogen atoms) are reactive and reactions with the first halogenoalkane produced occur, resulting in further substitution. Methane reacting with chlorine gives CH_3Cl initially, which reacts to give, successively, CH_2Cl_2, $CHCl_3$ and CCl_4.

$$CH_4 + Cl_2 \rightarrow CH_3Cl + HCl$$

> **Exam tip**
>
> A **substitution reaction** is where one atom or group replaces another in a molecule. There are always two reactants and two products.

then

$$CH_3Cl + Cl_2 \rightarrow CH_2Cl_2 + HCl \text{ etc}$$

Bromine reacts in a similar but slower reaction.

$$CH_4 + Br_2 \rightarrow CH_3Br + HBr$$

Iodine does not react with alkanes.

Direct halogenation is not a good way of making single compounds because a mixture of different halogenoalkanes, such as CH_3Cl, CH_2Cl_2, $CHCl_3$ and CCl_4 may be formed from methane. It is useful for making solvents, which can be mixtures.

Mechanism of radical substitution

The **mechanism** of a reaction is a way of representing how electrons are moved from the reactants to form the products. There are seven types of mechanism studied in the first and second years of the course. The halogenation of alkanes involves (**free**) **radical substitution**. It is also a **chain reaction**.

Curly arrows are used to show the movement of electrons. They must be drawn with precision. The tail of the arrow must come from the electron(s); the head must be where the electrons go. The movement of a single electron is shown by a half-headed arrow and that of an electron pair by a full-headed arrow.

The first step in the chlorination of methane produces chlorine radicals (atoms), the Cl–Cl bond having been broken by UV or sunlight. The single electron is shown as a dot. This type of bond breaking, in which one electron from the covalent bond goes to each of the fragments produced, is called **homolytic fission**.

The first **initiation step** produces the chlorine radicals (Figure 18):

Figure 18

The feature of a chain reaction is that the **propagation steps** that follow initiation produce one radical for every radical consumed. A chlorine radical attacks the methane molecule, producing a methyl radical and hydrogen chloride:

$$CH_4 + Cl\bullet \rightarrow HCl + \bullet CH_3$$

The methyl radical then attacks a chlorine molecule, to give chloromethane and a chlorine radical:

$$\bullet CH_3 + Cl_2 \rightarrow CH_3Cl + Cl\bullet$$

The chlorine radical attacks a methane molecule, and so the chain is propagated.

There are several possible **termination steps** where two radicals react to form a molecule, but not further radicals, thus ending the chain propagation:

$$\bullet CH_3 + \bullet CH_3 \rightarrow C_2H_6$$

$$\bullet CH_3 + Cl\bullet \rightarrow CH_3Cl$$

$$Cl\bullet + Cl\bullet \rightarrow Cl_2$$

The mechanism for the bromination of methane is similar.

Exam tip

A **radical** is an atom or group of atoms with an unpaired electron.

Exam tip

Remember that CH_3Cl and $Cl\bullet$ are not formed in the first propagation step.

Exam tip

The presence of ethane as a minor product is evidence for this mechanism.

Topic 6C Alkenes

Alkenes are a homologous series with the general formula C_nH_{2n}. An alkene also contains a C=C double bond. Because of this double bond, alkenes are **unsaturated** compounds. Cycloalkenes contain a ring of carbon atoms as well as a C=C group in the ring. An example is cyclohexene, C_6H_{10}.

The double bond consists of two bonds that are not identical. Head-on overlap of carbon orbitals gives a σ-bond, in which the electron density is coaxial with the C–C internuclear axis. There is also sideways overlap between p-orbitals on the carbon atoms, which gives a π-bond as the other part of the double bond. The σ-bond is shorter than in saturated compounds (e.g. ethane), so is somewhat stronger; the σ-bond component accounts for about two-thirds of the overall strength of the C=C bond.

Reaction mechanisms

Bond polarity and mechanism

The polarity of a bond determines the type of reagent that will attack it.

The C=C bond in alkenes and many other compounds is not polar, so it is not attacked by nucleophiles. It contains an accessible cloud of electron density in the π-bond, so it is attacked by electrophiles such as Br_2 or HBr. The result is **electrophilic addition**.

Alkanes are not polar and have no π-bonds with a high electron density. They are attacked only by radicals — reactive species that can break the strong σ-bonds. The result is **radical substitution**.

The carbon–halogen σ-bond in halogenoalkanes is polar. It has no area of high electron density, so the δ+ carbon atom is attacked by nucleophiles, resulting in **nucleophilic substitution**.

Drawing mechanisms

A drawing of a reaction mechanism is a static description of a dynamic process. By means of curly arrows, which show how electrons move, you are depicting the dynamics of how the electrons change their positions as the reaction proceeds. It is essential to have a picture in your mind of this dynamic process — do not simply regard the mechanism as a series of drawings, but rather as the starting and finishing frames of an animation.

Movement of a *single* electron is represented by half-headed arrows (having a *single* barb). Movement of a *pair* of electrons is shown by a full-headed arrow with a *twin* barb. Arrows should be drawn with some care because you are showing where electrons start from and where they end up.

Addition reactions of alkenes

Propene, $CH_3CH=CH_2$, is a good example to show in addition reactions since it is an unsymmetrical alkene, i.e. the substituents on each end of the C=C bond are not the same. This is significant in some reactions.

Knowledge check 7

Draw the structural formula of cyclohexene.

Exam tip

The type of *mechanism* is free radical, electrophilic or nucleophilic. The type of *reaction* is addition, substitution etc.

Exam tip

An **addition reaction** is where two substances form a single substance.

Reaction with hydrogen

Propene reacts with hydrogen at 150°C with a nickel catalyst, to give propane:

$$CH_3CH=CH_2 + H_2 \rightarrow CH_3CH_2CH_3$$

The type of mechanism is free radical. The catalyst splits up hydrogen molecules into atoms which then add on to the adsorbed alkene. This particular reaction is not useful industrially — alkenes are far more valuable than alkanes, since they are more reactive. However, it is the basis of margarine manufacture. Vegetable oils are polyunsaturated triesters and reduction of some of the C=C bonds produces a harder fat which is then mixed with additives such as vitamins and emulsifiers and sold as margarine.

Reaction with halogens

Propene reacts with chlorine or bromine in the gas phase, or in an inert solvent (e.g. CCl_4), at room temperature:

$$CH_3CH=CH_2 + Br_2 \rightarrow CH_3CHBrCH_2Br$$

This type of reaction is addition. The product is 1,2-dibromopropane. The brown bromine is decolourised.

The reaction of bromine *water* (aqueous bromine) with alkenes is used to test for the presence of a C=C bond — the orange bromine water is decolourised in a positive test. HOBr is added across the double bond, not Br_2:

$$CH_3CH=CH_2 + HOBr \rightarrow CH_3CHBrCH_2OH$$

Mechanism of addition of bromine

This reaction is an example of **electrophilic addition**, see Figure 19.

An **electrophile** is a species which forms a bond with an electron-rich site in a molecule. It does so by accepting a **pair** of electrons from that site forming a new covalent bond.

The electrophile causes **heterolytic** fission of a covalent bond in the organic molecule. The result is the formation of a carbocation (where a carbon atom is +) as well as a new covalent bond with the electrophile. The electrophile is a Br atom in the Br_2 molecule.

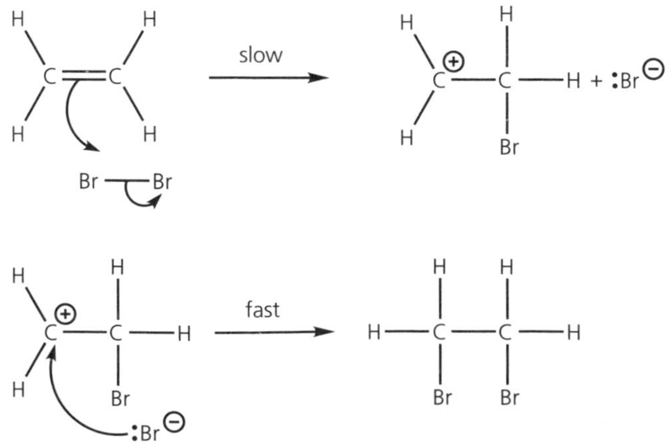

> **Exam tip**
>
> The HOBr is formed by the reaction of bromine with water:
>
> $Br_2 + H_2O \rightarrow HOBr + HBr$

Figure 19

Reaction with hydrogen halides

Propene reacts with hydrogen bromide in the gas phase, or in an inert solvent, at room temperature to form 2-bromopropane (see below) as the major product, together with some 1-bromopropane. The mechanism is electrophilic addition.

$$CH_3CH=CH_2 + HBr \rightarrow CH_3CHBrCH_3$$

The hydrogen goes to the carbon atom in the double bond that already has the most hydrogen atoms on it. This is based on Markovnikoff's rule; this 'rule' is *not* an explanation and there are exceptions to it. It is, however, a useful guide.

Hydrogen chloride and hydrogen iodide react with propene in a similar manner to hydrogen bromide.

Mechanism of addition of hydrogen bromide

Figure 20 shows the mechanism of electrophilic addition of hydrogen bromide with ethene. The electrophile is the $\delta+$ H atom in HBr.

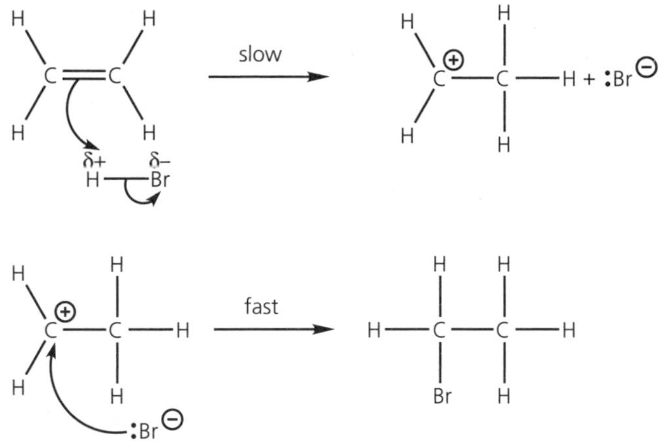

Figure 20

Figure 21 shows the two-step mechanism with propene.

Step 1

Secondary carbocation — major

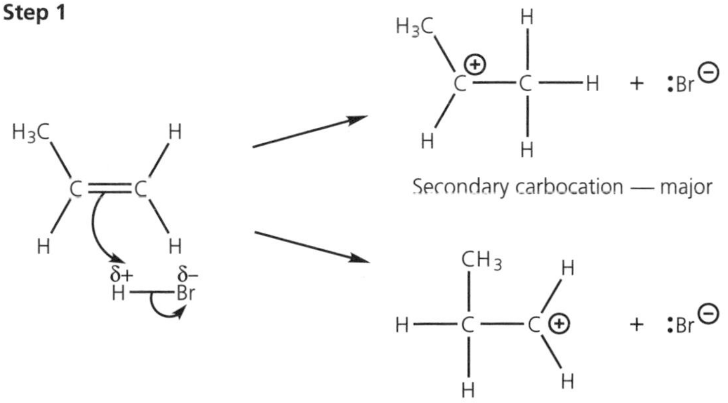

Primary carbocation — minor. In the secondary carbocation intermediate, the C^+ is stabilised by the electron-pushing effect of a CH_3 group. Thus it is more likely to be formed than the primary carbocation, which does not have a CH_3 group attached to the C^+.

> **Exam tip**
>
> Remember to add the partial (δ) charges to the electrophile.

> **Exam tip**
>
> The order of stability of carbocations is: tertiary > secondary > primary

Step 2

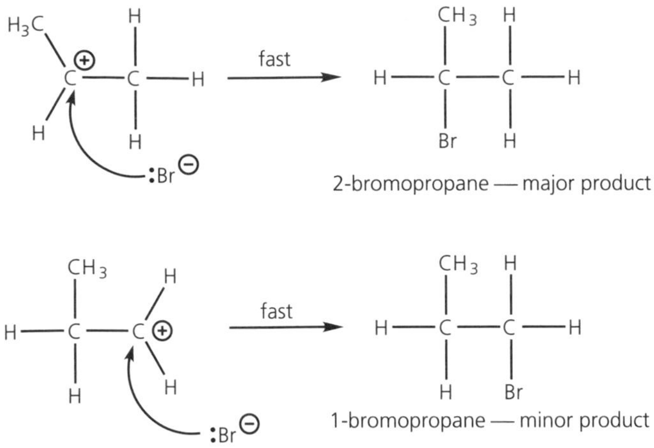

2-bromopropane — major product

1-bromopropane — minor product

Figure 21

Reaction with steam

If an alkene is passed with steam under pressure over heated phosphoric acid catalyst an addition reaction takes place.

With ethene the product is ethanol. With propene the major product is propan-2-ol.

$$CH_2=CH_2 + H_2O \rightarrow CH_3CH_2OH$$

$$CH_3CH=CH_2 + H_2O \rightarrow CH_3CH(OH)CH_3$$

Reaction with potassium manganate(VII) solution

Propene reacts with aqueous potassium manganate(VII) solution to give propane-1,2-diol:

$$CH_3CH=CH_2 + [O] + H_2O \rightarrow CH_3CH(OH)CH_2OH$$

The precise nature of this reaction is complex, so a balanced equation is not written. The purple $KMnO_4$ solution is usually converted to a brown sludge of MnO_2.

Polymerisation of alkenes

Polyalkenes are made by radical addition reactions. A radical initiator such as a peroxide or oxygen is heated with the alkene. The first step is dissociation of the initiator into radicals (Figure 22):

Figure 22

Exam tip

In strongly alkaline solution, the MnO_4^- ions are reduced to green MnO_4^{2-} ions and in strongly acid solution to the colourless Mn^{2+} ions. The organic equations are the same as with neutral $KMnO_4$ (potassium manganate).

The radical then attacks the alkene to form a new radical (Figure 23), and so on:

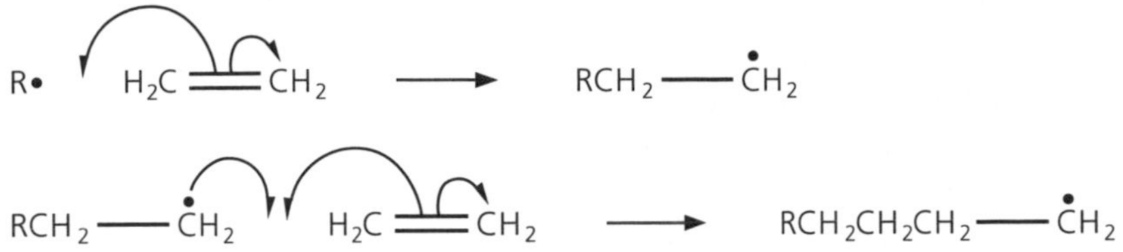

$$R\bullet \quad H_2C = CH_2 \longrightarrow RCH_2 \text{---} \dot{C}H_2$$

$$RCH_2 \text{---} \dot{C}H_2 \quad H_2C = CH_2 \longrightarrow RCH_2CH_2CH_2 \text{---} \dot{C}H_2$$

Figure 23

Because radicals are very reactive, they can attack growing chains at any point, forming branches and cross-links between the chains. The reaction stops when two radicals combine. Polymers contain molecules with a wide variety of chain lengths, so they soften over a range of temperatures rather than melting sharply.

Alkene monomers can contain substituent groups other than hydrogen. These groups *never* form part of the central two-carbon chain. This is illustrated in Table 5.

Monomer	Polymer	Name
$CH_2=CH_2$	$-\!\!\left(CH_2 \text{---} CH_2\right)_{\!n}$	Poly(ethene)
$CH_2=CHCH_3$	$-\!\!\left(CH_2 \text{---} \underset{\underset{CH_3}{\mid}}{CH}\right)_{\!n}$	Poly(propene)
$CH_2=CHCl$	$-\!\!\left(CH_2 \text{---} \underset{\underset{Cl}{\mid}}{CH}\right)_{\!n}$	Poly(chloroethene), PVC
$CF_2=CF_2$	$-\!\!\left(CF_2 \text{---} CF_2\right)_{\!n}$	Poly(tetrafluoroethene), PTFE

Table 5

Polymers and resources

Polymers are made from oil and need energy for their manufacture. They are unreactive. Burning polymers can gives rise to toxic fumes. With specially designed factories the toxic fumes can be removed and the energy released can be turned into electricity.

Most polymers can be recycled, but recycling has energy costs in terms of transport and reprocessing. Sorting the different polymer varieties is expensive because there are many different varieties that are not always easily distinguishable. On the other hand, polymers are durable and take up landfill space, are polluting and can harm wildlife.

Exam tip

The repeat unit of a polymer made by addition will contain a chain of two carbon atoms. Thus the repeat unit of (poly)propene is *not* $-(CH_2\text{-}CH_2\text{-}CH_2)-$.

Knowledge check 8

Draw the repeat unit from the polymerisation of propenoic acid, $CH_2=CHCOOH$.

Waste polymers can also be used as a feedstock for cracking, thus producing alkenes (for making new polymers) and alkanes (used as fuel).

Wrapping food in polymers improves its keeping quality and has, therefore, reduced food waste and the incidence of food-related illness. In some cases, paper products could be used instead of polymers — unlike oil, paper is a renewable resource. However, the energy consumption in paper manufacture is significant. It also requires considerable quantities of water, the supply of which can be more of a problem than finding suitable trees — most of these are grown as a crop. Paper is made from the timber of conifers. The soil in which conifers are grown emits nitrogen oxides into the atmosphere and can contribute a greater volume of nitrogen oxides to the locality than motor vehicles do. So, paper cups are not necessarily better than poly(styrene) ones.

Biodegradable polymers, such as poly(3-hydroxybutanoic acid), are more environmentally friendly as they are broken down in the soil or waste dumps. However, they are expensive and lack tensile strength.

Exam tip

Recycling plastics is *not* carbon neutral. Using paper from trees instead of plastics is not always better for the environment.

Summary

After studying this topic, you should:
- be able to give the systematic names of alkanes and alkenes and of the products of their reactions
- be able to draw and name structural and *E/Z* (geometric) isomers of alkanes and alkenes
- know the equations and conditions for the reactions of:
  - alkanes with air (oxygen) and chlorine
  - alkenes with hydrogen, halogens, hydrogen halides, steam and potassium manganate(VII)

- be able to deduce the repeat units of polymers given the formula of the monomer
- draw the mechanism of the free radical substitution reaction of alkanes
- draw the mechanism of electrophilic addition reactions of alkenes

Topic 6D Halogenoalkanes

Halogenoalkanes are compounds in which one or more of the hydrogen atoms of an alkane have been substituted by halogen atoms. The simplest have the general formula $C_nH_{2n+1}X$ where X is Cl, Br or I. There are three general types of halogenoalkane — primary secondary and tertiary.

- Primary have just one (or no) carbon atom attached to the C–X group.
- Secondary have two carbon atoms attached to the C–X group.
- Tertiary have three carbon atoms attached.

Exam tip

Primary halogenoalkanes have a –CH_2X group and secondary halogenoalkanes a >CHX group. Tertiary halogenalkanes have no H atom on the carbon with the halogen atom.

The examples in Table 6 are isomers of C_4H_9Br.

Primary	Secondary	Tertiary
1-bromobutane ($CH_3CH_2CH_2CH_2Br$)	2-bromobutane ($CH_3CH_2CH(Br)CH_3$)	2-bromo-2-methylpropane ($(CH_3)_3CBr$)

Table 6 Isomers of halogenoalkane C_4H_9Br

Nucleophilic substitution reactions

A nucleophile is a species with a lone pair of electrons that it uses to form a covalent bond with a δ+ atom in another molecule. Substitution is where one atom or group is replaced by another.

In the reactions of halogenoalkanes the δ+ carbon atom is attacked by the nucleophile. The reactions are similar for chloro-, bromo- and iodoalkanes. The rates of reaction increase in the order Cl < Br < I. This is because the C–I bond is the weakest, making the activation energy (see page 59) for the reaction smaller.

The rate of reaction for a given halogen in halogenoalkanes decreases in the order tertiary > secondary > primary. Thus 2-chloro-2-methylpropane reacts faster than 2-chlorobutane which reacts faster than 1-chlorobutane.

Reaction with sodium hydroxide in aqueous solution

Halogenoalkanes heated under reflux with aqueous NaOH (or KOH) give mainly the alcohol, in a nucleophilic substitution reaction:

$$CH_3CH_2CH_2Br + NaOH \rightarrow CH_3CH_2CH_2OH + NaBr$$

> **Exam tip**
>
> The nucleophile could be a negative ion, such as OH^- or CN^-, or a neutral molecule, such as H_2O or NH_3.

The ionic equation for this reaction is:

$$CH_3CH_2CH_2Br(l) + OH^-(aq) \rightarrow CH_3CH_2CH_2OH(aq) + Br^-(aq)$$

Some ethanol is often added to improve the miscibility of the reagents.

Mechanism of the reaction with OH⁻ ions

There are two types of nucleophilic substitution when hydroxide ions attack a halogenoalkane. In both, the nucleophile is the OH^- which bonds using its lone pair of electrons.

S_N1

The S_N1 reaction (substitution nucleophilic unimolecular) occurs with tertiary and to some extent with secondary halogenoalkanes. It is a fast reaction with tertiary halogenoalkanes (Figure 24). These compounds are able to generate carbocations that are more stable than primary carbocations. The mechanism has two steps. The first step is the rate-determining (slow) step. If the starting halogenoalkane is chiral (see the student guide covering topics 16–19 in this series), the product mixture has equal amounts of the two optical isomers of the product and so is not optically active.

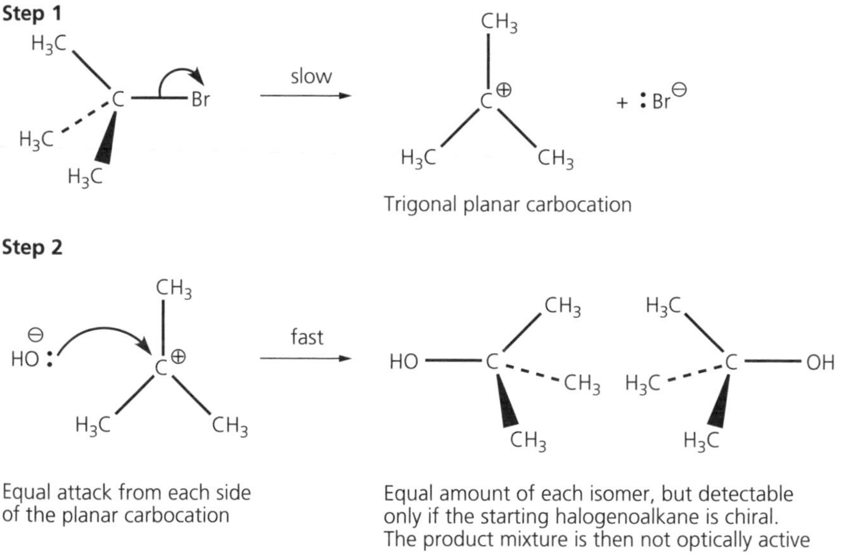

Step 1

Trigonal planar carbocation

Step 2

Equal attack from each side of the planar carbocation

Equal amount of each isomer, but detectable only if the starting halogenoalkane is chiral. The product mixture is then not optically active

Figure 24

S_N2

The S_N2 reaction (substitution nucleophilic bimolecular) occurs with primary halogenoalkanes, such as bromoethane, and is a fairly slow reaction (Figure 25). Heterolytic fission of the carbon–halogen bond (as in the S_N1 reaction above) would yield a primary carbocation, so the energetically more favoured mechanism has only one step.

Exam tip

Always write this mechanism in two steps with the carbocation as a reactant in the second step.

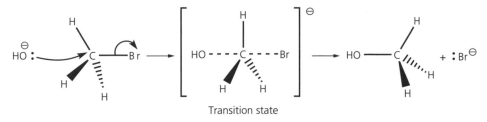

Transition state

Figure 25

A chiral halogenoalkane gives an alcohol in which the arrangement of the substituent groups is inverted (see the student guide covering topics 16–19 in this series).

In synthetic processes where the stereochemistry of the product is critical (e.g. in the manufacture of pharmaceuticals), a careful mechanistic analysis of the intermediate stages is necessary for the overall synthetic pathway to be successful.

Reaction with water containing silver ions

When halogenoalkanes are shaken with aqueous silver nitrate, a precipitate of the silver halide is observed. In this reaction the water is the nucleophile. Hydrolysis first produces halide ions, which then react with silver ions to form a precipitate of silver halide:

$$R–X(l) + H_2O(l) \rightarrow ROH(aq) + H^+(aq) + X^-(aq)$$

then:

$$Ag^+(aq) + X^-(aq) \rightarrow AgX(s)$$

Core practical 4: Investigation of the rates of hydrolysis of some halogenoalkanes

The relative rates of the hydrolysis of different halogenoalkanes can be found by measuring the time taken for the precipitate of silver chloride to appear. The procedure is:

1 Add four drops of the halogenoalkane to a test tube and a few drops of ethanol to increase solubility.
2 Now add $2\,cm^3$ of aqueous silver nitrate, stopper the tube with a cork and invert and shake.
3 Note the time for a precipitate to appear.
4 Repeat with the same quantities using a different halogenoalkane.

It is observed that the precipitate will first appear with iodoalkanes, then with similar bromoalkanes and slowest with chloroalkanes. Likewise, the precipitate appears first with tertiary halogenoalkanes, then with secondary and slowest of all with primary.

Exam tip

Even though iodoalkanes are less polar than bromo- or chloroalkanes, they are hydrolysed faster because the C–I bond is the weakest.

Knowledge check 9

Which will react by nucleophilic substitution with hydroxide ions faster:

a 2-bromo-2-methylpropane or 1-bromo-2-methylpropane?

b 2-bromopropane or 2-chloropropane?

Exam tip

Do not try to write an equation between the halogenoalkane and silver ions. It reacts with water first.

Reaction with potassium cyanide

When a halogenoalkane is warmed with potassium cyanide dissolved in a mixture of ethanol and water, a nitrile is formed in a nucleophilic substitution reaction.

$$CH_3CH_2CH_2Br + KCN \rightarrow CH_3CH_2CH_2CN + KBr$$

The ionic equation is

$$CH_3CH_2CH_2Br + {}^-CN \rightarrow CH_3CH_2CH_2CN + Br^-$$

Mechanism of the reaction with cyanide ions

This is identical to the reaction of a halogenoalkane with hydroxide ions except that it is the lone pair of electrons on the carbon of the cyanide ion that attacks the $\delta+$ carbon atom in the halogenoalkane.

Reaction with ammonia

Halogenoalkanes heated in a sealed tube with concentrated ammonia in ethanol give a mixture of amines. The primary amine (RNH_2) can be made the major product by using excess ammonia. The reaction is a nucleophilic substitution:

$$CH_3CH_2CH_2CH_2Br + 2NH_3 \rightarrow CH_3CH_2CH_2CH_2NH_2 + NH_4{}^+Br^-$$

A mixture is obtained because the amine produced is also a nucleophile and attacks any unchanged halogenoalkane.

Mechanism

The nitrogen atom in NH_3 has a lone pair of electrons and thus is a nucleophile. This attacks the $\delta+$ carbon atom of the halogenoalkane. Figure 26 shows the mechanism with a primary halogenoalkane:

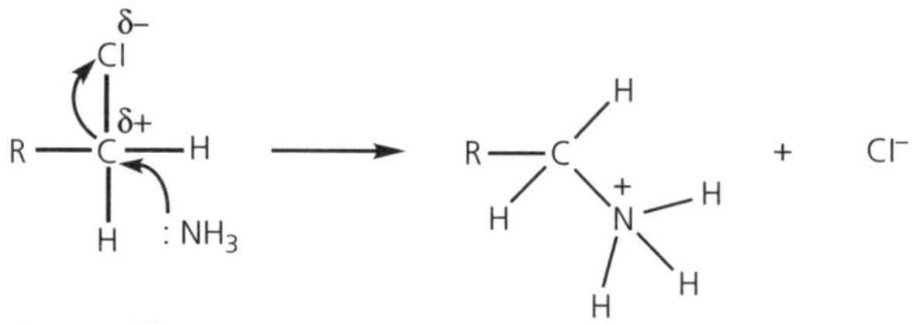

Figure 26

H^+ is then lost from the $^+NH_3$ group (Figure 27).

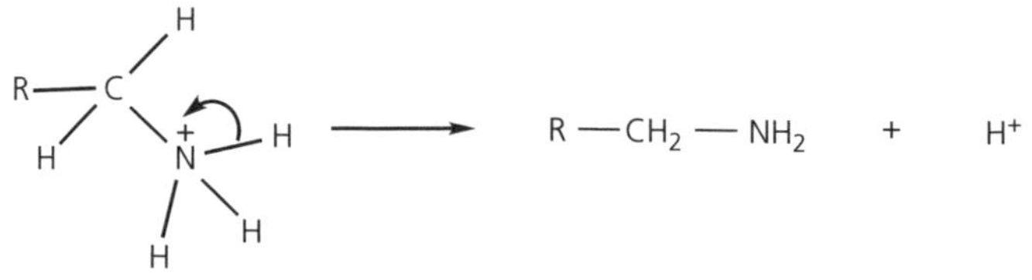

Figure 27

Finally the H^+ ions react with excess ammonia

$$H^+ + NH_3 \rightarrow NH_4{}^+$$

Exam tip

It is a common error to have the curly arrow starting from the nitrogen of the cyanide ion, especially if it is written CN^-. Remember that the new bond is formed between the C in the halogenoalkane and the C in the CN.

Exam tip

The reaction of water with halogenoalkanes has a similar mechanism. Step 1 is the nucleophilic attack by a lone pair of electrons on the oxygen, followed by loss of H^+ in step 2.

Elimination reaction

In this reaction the halogen atom and a hydrogen atom on an adjacent carbon atom are removed and a carbon–carbon double bond is formed.

Reaction with potassium hydroxide in ethanolic solution

With ethanolic KOH, halogenoalkanes eliminate HX (where X is a halogen) to give an alkene.

$$CH_3CH_2CH_2CH_2Br + KOH \rightarrow CH_3CH_2CH=CH_2 + KBr + H_2O$$

Identifying the halogen atom in a halogenoalkane

As the hydrolysis of some halogenoalkanes is very slow, the most reliable method is as follows:

1 Heat the halogenoalkane with sodium hydroxide solution to hydrolyse the halogenoalkane to the alcohol and sodium halide:

$$CH_3Br + NaOH \rightarrow CH_3OH + Na^+ + Br^-$$

2 Acidify the solution with nitric acid (test with litmus).

3 Add silver nitrate solution.

A white precipitate soluble in dilute ammonia indicates the formation of a chloride; a cream precipitate soluble in concentrated ammonia indicates a bromide; and a yellow precipitate insoluble in ammonia indicates an iodide.

Apart from the initial hydrolysis, these reactions are the standard tests for halide ions.

Topic 6E Alcohols

Alcohols are organic compounds containing one (or more) hydroxyl groups (–OH) as the functional group. The **general formula** of simple alcohols is $C_nH_{2n+1}OH$

There are three types of alcohol that have one –OH group. Each type is shown in Table 7 on pages 32–33, together with the names and different representations of the isomers of the first four members of the alcohol homologous series. Skeletal formulae are not generally used where a compound has fewer than three carbon atoms. In each case, R, R′ and R″ are organic groups that can be different or the same. For primary alcohols alone, R can be H (giving methanol).

> **Exam tip**
>
> Elimination is much faster with tertiary halogenoalkanes which is why, if substitution is required, they should not be warmed with aqueous alkali.

> **Knowledge check 10**
>
> Describe the test and result that would show the presence of an alkene.

> **Exam tip**
>
> Note that –OH is not 'hydroxide'.

Content Guidance

Exam tip

Primary alcohols have a $-CH_2OH$ group and secondary alcohols have a >CH(OH) group. Tertiary alcohols have no H atom on the carbon with the $-OH$ group.

Primary alcohols	Secondary alcohols	Tertiary alcohols
Methanol (CH_3OH) 	—	—
Ethanol (CH_3CH_2OH) 	—	—
Propan-1-ol ($CH_3CH_2CH_2OH$) 	Propan-2-ol ($CH_3CH(OH)CH_3$) 	—

Primary alcohols	Secondary alcohols	Tertiary alcohols
Butan-1-ol ($CH_3CH_2CH_2CH_2OH$)	Butan-2-ol ($CH_3CH_2CH(OH)CH_3$)	2-methylpropan-2-ol ($(CH_3)_3COH$)

and

2-methylpropan-1-ol ($CH_3CH(CH_3)CH_2OH$)

Table 7 Primary, secondary and tertiary alcohols

Reactions of alcohols

Combustion

All alcohols burn in air to give carbon dioxide and water:

$$CH_3CH_2OH + 3O_2 \rightarrow 2CO_2 + 3H_2O$$

Because they contain an oxygen atom, they burn with a cleaner flame than hydrocarbons do. Alcohols become less easy to burn as the carbon chain gets longer because they become less volatile.

The general equation for the combustion of alcohols is:

$$C_nH_{2n+1}OH + (3n/2)O_2 \rightarrow nCO_2 + (n + 1)H_2O$$

Ethanol can be produced by the fermentation of sugar cane and is added to petrol for motor fuel in Brazil. The ethanol is therefore a renewable fuel, but any land given over to growing fuel cannot be used for growing food (see page 19).

Oxidation with potassium dichromate(VI) in sulfuric acid

In general, full equations are not written for these oxidation reactions. The oxidising agent (potassium dichromate(VI) in dilute sulfuric acid) is represented by [O].

Primary alcohols react to give an aldehyde, which, if not removed from the reaction system, oxidises further to yield a carboxylic acid.

$$CH_3CH_2OH + [O] \rightarrow CH_3CHO + H_2O$$

$$CH_3CHO + [O] \rightarrow CH_3COOH$$

The mixture changes colour from orange to green.

Secondary alcohols react to give ketones. The ketones are not oxidised further under these conditions, so the mixture can be heated under reflux.

$$CH_3CH(OH)CH_3 + [O] \rightarrow CH_3COCH_3 + H_2O$$

Tertiary alcohols do not react under these conditions.

Core practical 5: Preparation of ethanal by oxidation of ethanol

This standard preparation produces ethanal in aqueous solution. It is invariably contaminated with a small amount of ethanoic acid.

Step	Reason
(1) Place some water in a round-bottomed flask and add some concentrated sulfuric acid *slowly*, while shaking. Add some anti-bumping granules.	**Why must the sulfuric acid be added slowly, while shaking?** The reaction between sulfuric acid and water is dangerously exothermic. If mixed rapidly so much heat will be produced that the mixture will boil and spray out hot acidic solution. **What are anti-bumping granules and what is their purpose?** The granules are silicon dioxide. They prevent the formation of large gas bubbles that cause 'bumping'.
(2) Add the flask containing the acid into the assembled distillation apparatus. The still head is fitted with a tap funnel instead of a thermometer. The receiving flask is placed in an ice-water bath.	**Why is the receiving flask surrounded by ice-water?** The boiling temperature of ethanal is 21°C, so the cooling reduces loss of ethanal by evaporation.
(3) Dissolve some potassium dichromate(VI) in water and add the ethanol. Place the mixture in the tap funnel. Heat the contents of the round-bottomed flask to almost boiling and turn off the flame.	**Why isn't the ethanol oxidised by the dichromate(VI) solution?** The oxidation requires H^+ ions, and the mixture has not yet been acidified.

→

Exam tip

Make sure that you can draw the diagrams of the apparatus for the preparation of ethanal and propanone.

Knowledge check 11

What would you observe if 2-methylpropan-1-ol and 2-methylpropan-2-ol were each heated with acidified dichromate(VI) ions?

Knowledge check 12

Why does ethanal have a much lower boiling temperature than ethanol?

Step	Reason
(4) *Slowly* add the ethanol/dichromate(VI) solution into the flask. A vigorous reaction takes place and the ethanal is immediately boiled off and condensed in the condenser. The solution turns green as Cr^{3+} ions are formed. The cooled receiving flask contains a mixture of ethanal and water plus a little ethanoic acid and unreacted ethanol.	**Why is the ethanol/dichromate(VI) mixture added slowly to the hot acid?** Rapid addition would lead to an excess of oxidising agent and some of the ethanal formed would be oxidised to ethanoic acid.

If the ethanol and acidified dichromate(VI) mixture is heated under reflux, the ethanal initially formed is completely oxidised to ethanoic acid. This can then be distilled out of the product mixture.

Knowledge check 13

Suggest two reasons why the industrial manufacture of ethanal uses vapour-phase oxidation of ethanol by air over a heated catalyst.

Reaction with halogenating agents

Alcohols can be converted into halogenoalkanes in several ways, which work for all three types of alcohol.

Using phosphorus pentachloride

Solid phosphorus pentachloride (PCl_5) reacts readily with alcohols at room temperature to give the chloroalkane and HCl gas, which is emitted as steamy fumes:

$$CH_3CH_2OH + PCl_5 \rightarrow CH_3CH_2Cl + HCl + POCl_3$$

Test for the –OH group

Reaction with PCl_5 can be used as a test for presence of an –OH (hydroxyl) group. On addition of phosphorus pentachloride, steamy fumes (of HCl) are produced. When a glass rod dipped in concentrated ammonia is held in the fumes, white smoke (of NH_4Cl) is observed — this shows the presence of an OH group. This test is not specific to alcohols since the –OH groups in carboxylic acids give the same reaction, but these acids would also affect indicators, which alcohols do not.

Using a mixture of sodium bromide and 50% sulfuric acid

The alcohol is heated under reflux with sodium bromide and 50% sulfuric acid.

$$CH_3CH_2OH + NaBr + H_2SO_4 \rightarrow CH_3CH_2Br + NaHSO_4 + H_2O$$

Using a mixture of iodine and moist red phosphorus

Warming alcohols with iodine and moist red phosphorus produces iodoalkanes. The first reaction yields phosphorus triiodide, which reacts with the alcohol.

$$P_4 + 6I_2 \rightarrow 4PI_3$$

$$3CH_3CH_2OH + PI_3 \rightarrow 3CH_3CH_2I + H_3PO_3$$

Exam tip

This method cannot be used for iodoalkanes as sulfuric acid will oxidise the iodide ions to iodine.

Reaction of a tertiary alcohol with concentrated hydrochloric acid

Tertiary alcohols are easily protonated. The intermediate formed then loses water forming a tertiary carbocation that then picks up a Cl^- ion. The overall reaction with 2-methylpropan-2-ol is:

$$(CH_3)_2C(OH)CH_3 + HCl \rightarrow (CH_3)_2CClCH_3 + H_2O$$

Exam tip

This reaction works with HBr and also with HI.

Core practical 6: Chlorination of 2-methylpropan-2-ol by concentrated hydrochloric acid

The procedure is:
- Place $15\,cm^3$ of the tertiary alcohol and $50\,cm^3$ of concentrated hydrochloric acid in a conical flask.
- Put a rubber bung in the flask and shake the mixture thoroughly for about 20 minutes, releasing any pressure every 2 minutes. By now there should be two visible layers.
- Add some anhydrous calcium chloride and shake. Transfer the mixture to a separating funnel and discard the lower aqueous layer.
- Remove any acid from the organic layer by shaking with sodium hydrogencarbonate solution, frequently releasing the pressure due to carbon dioxide build-up.
- Discard the aqueous layer and run the organic layer into a conical flask. Add some lumps of anhydrous sodium sulfate (a drying agent) and leave until the liquid becomes clear.
- Finally transfer the liquid to a distillation flask. Set this up for distillation using a water condenser and distil off the fraction that boils between 50°C and 53°C.

Exam tip

This method illustrates the uses of a separating funnel, distillation and drying.

Summary

After studying this topic, you should be able to:
- distinguish between, name and write the formulae of primary, secondary and tertiary alcohols and halogenoalkanes
- describe and write equations for the reactions of alcohols with air (combustion), halogenating agents and acidified potassium dichromate(VI)
- describe the preparation of an aldehyde or ketone from an alcohol and the preparation of a halogenoalkane from an alcohol and explain why certain procedures are used
- describe an experiment to show the relative reactivity of different halogenoalkanes
- write equations for the reactions of halogenoalkanes with aqueous alkali, aqueous silver nitrate, potassium cyanide, ammonia and alcoholic potassium hydroxide
- draw the mechanisms for the reactions of halogenoalkanes with OH^- ions, CN^- ions and with ammonia

■Topic 7 Modern analytical techniques I

Topic 7A Mass spectrometry

The mass spectroscope was invented by Francis W. Aston, who received the Nobel prize in chemistry in 1922 for using his instrument to detect isotopes. The mass spectrometer is the modern equivalent and is described in the student guide covering topics 1–5 in this series.

When the vaporised organic sample passes into the ionisation chamber of a mass spectrometer, it is bombarded by a stream of electrons. These electrons have a high enough energy to knock an electron off an organic molecule to form a positive ion. This ion is called the **molecular ion**, symbol M^+. Molecular ions tend to be unstable and some of them break into smaller fragments. For example, one of the ways the molecular ion of ethanol breaks down is:

$$CH_3CH_2OH^+ \quad \rightarrow \quad CH_3\bullet \quad + \quad CH_2OH^+$$

molecular ion   radical   fragment ion

Exam tip

When an ion breaks down, a radical and another (positive) ion is formed.

- The machine prints out a bar graph, which shows the abundance of each ion plotted against m/z. This is called the **mass spectrum**.
- The technique is sensitive and fragmentation patterns are characteristic, so they can be used to identify compounds.
- The height of the peak from the most abundant ion is scaled to a value of 100.

Examples of mass spectra

Propanal and propanone

Figure 28 shows the mass spectrum of propanal.

Knowledge check 14

Identify the species that would be present in the mass spectrum of 2-methylbutane and not in its isomer 2,2-dimethylpropane.

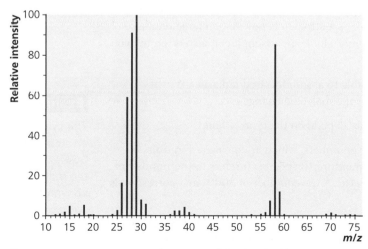

Figure 28 Mass spectrum of propanal (CH_3CH_2CHO)

The most significant peaks in the mass spectrum of propanal are listed below.

- The peak at $m/z = 58$ is the molecular ion peak arising from $CH_3CH_2CHO^+$. Not all compounds give a molecular ion peak. There is sometimes a small peak one unit higher in mass arising from the presence of the ^{13}C isotope in the molecule.
- The peak at $m/z = 57$ is due to $CH_3CH_2CO^+$.
- The peak at $m/z = 29$ is due to CHO^+ and $CH_3CH_2^+$.
- The peak at $m/z = 15$ is due to CH_3^+.

Figure 29 shows the mass spectrum of propanone.

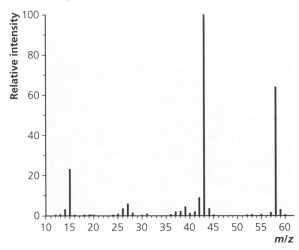

Figure 29 Mass spectrum of propanone (CH_3COCH_3)

- The peak at $m/z = 58$ is the molecular ion peak from $CH_3COCH_3^+$.
- There is a small peak at $m/z = 59$ from molecules containing ^{13}C.
- The peak at $m/z = 43$ is from CH_3CO^+ formed by loss of CH_3.
- The peak at $m/z = 15$ is from CH_3^+.

Topic 7B Infrared (IR) spectroscopy

Infrared (IR) spectroscopy is a powerful tool for discovering which functional groups are present in a compound. Since most IR spectrometers now have a large database of known compounds installed in their memory for comparison, it can also often tell you what a substance is.

Only compounds with polar bonds are able to absorb infrared radiation. As they absorb infrared radiation, the polarity of the molecule changes.

The absorption frequencies of a molecule depend on the type of bond.

The horizontal scale is calibrated in wavenumbers, which are measured in units of cm^{-1}. This unit of cm^{-1} is used only to measure IR spectra. It refers to the number of wavelengths of the radiation per centimetre. A wavenumber of $3000\,cm^{-1}$ corresponds to a frequency of about $10^{14}\,Hz$.

The spectrum from about $1400\,cm^{-1}$ to $600\,cm^{-1}$ is called the **fingerprint region** and corresponds to bending vibrations of the molecule. Its complexity makes it useful for comparison between an unknown spectrum and a reference database since each compound has a unique fingerprint.

Exam tip

It is a common error to omit the + charge when identifying a peak.

Exam tip

Propanone does not have a peak at $m/z = 29$ because the molecule cannot fragment to give either CHO^+ or $CH_3CH_2^+$.

Exam tip

The peak at $m/z = 43$ is absent in propanal, because the molecule cannot fragment to give a CH_3CO^+ or a $C_3H_7^+$ ion.

Knowledge check 15

Identify the species that gives a peak at $m/z = 31$ in the mass spectrum of propan-1-ol.

Exam tip

The symmetrical stretching of linear CO_2 is not IR active, as the molecule remains non-polar. The bending and asymmetrical stretching cause a change in dipole and so it absorbs IR radiation.

The vertical axis is calibrated as percentage transmittance, so IR absorptions appear as troughs or dips in the spectrum. They are nevertheless usually referred to as peaks.

Types of bond stretch

C–H stretch

Figure 30 shows the IR spectrum of pentane.

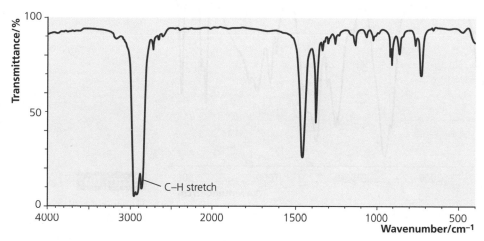

Figure 30 IR spectrum of pentane ($CH_3CH_2CH_2CH_2CH_3$)

The C–H stretch in alkanes occurs at approximately $3000–2800\,cm^{-1}$.

In alkenes it is around $3100–3000\,cm^{-1}$.

In aldehydes it is around $2900–2700\,cm^{-1}$.

O–H stretch

Figure 31 shows the IR spectrum of ethanol.

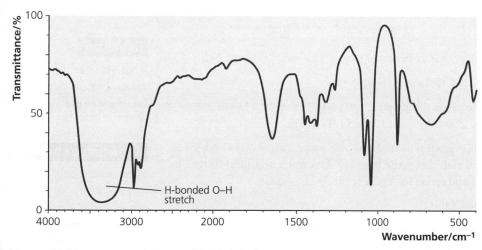

Figure 31 IR spectrum of ethanol (CH_3CH_2OH)

The hydrogen-bonded O–H stretch in alcohols occurs at about 3800–3200 cm^{-1}. The frequency at which the O–H bond vibrates depends on how tightly it is hydrogen-bonded to another molecule. This differs between molecules so the absorption peak is broad. If an alcohol is diluted with a solvent that does not form hydrogen bonds, the peak narrows and moves to about 3400 cm^{-1}.

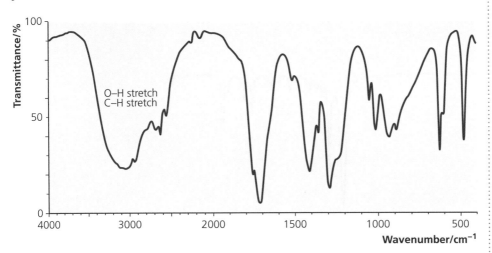

Figure 32 IR spectrum of ethanoic acid (CH$_3$COOH)

The O–H stretch in carboxylic acids occurs at about 3300–2500 cm^{-1}. The hydroxyl group is hydrogen-bonded forming a dimer (two molecules joined together, as shown in Figure 33). The O–H absorption is broad and lies over the C–H stretch.

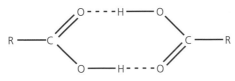

Figure 33

C=O stretch

The C=O stretch in aldehydes occurs between 1740 and 1720 cm^{-1} (Figure 34).

In ketones it is between 1725 and 1710 cm^{-1} (Figure 35).

The C=O stretch in carboxylic acids is between 1700 and 1725 cm^{-1}. They also have a broad peak between 3300 and 2500 cm^{-1} due to the O–H group.

The C=O stretching vibration in both propanal and propanone (Figures 34 and 35) is just above 1700 cm^{-1}. Propanal also shows the broad O–H stretch around 3400 cm^{-1}. This is due to the existence of the following equilibrium in the liquid:

$$CH_3CH_2CHO \rightleftharpoons CH_3CH=CHOH$$

The hydrogen-bonded N–H stretch (Figure 36) in amines is broad because of hydrogen bonding and occurs between 3500 and 3300 cm^{-1}.

Exam tip

The O–H stretch can be used to distinguish between alcohols and acids as they are at different wavenumbers.

Exam tip

A list of absorption wavenumbers can be found in the data booklet.

Knowledge check 16

What peak would be present in the IR spectrum of propylamine, CH$_3$CH$_2$CH$_2$NH$_2$ but not in its isomer trimethylamine (CH$_3$)$_3$N?

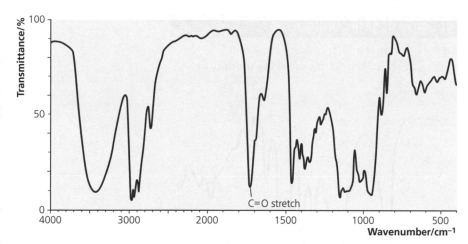

Figure 34 IR spectrum of propanal (CH_3CH_2CHO)

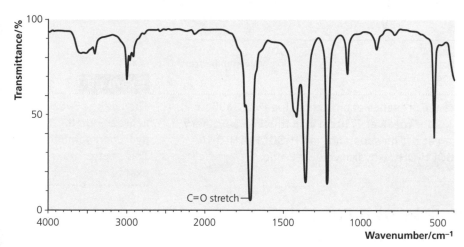

Figure 35 IR spectrum of propanone (CH_3COCH_3)

N–H stretch

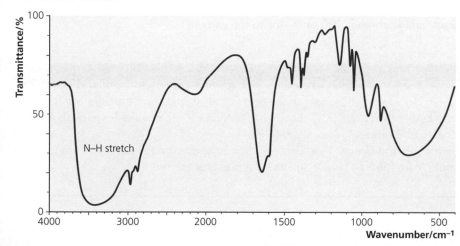

Figure 36 IR spectrum of ethylamine ($CH_3CH_2NH_2$)

Core practical 7: Analysis of some organic unknowns

An organic compound of formula $C_4H_8O_2$ has the IR spectrum shown in Figure 37. It can be oxidised to form a carboxylic acid. Suggest the identity of the compound.

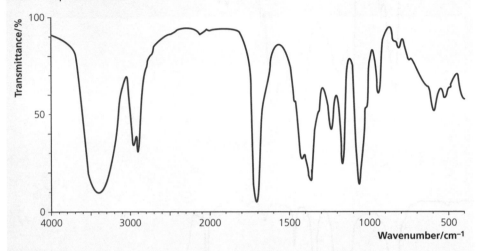

Figure 37

Answer: The spectrum indicates the presence of an alcohol (peak at $3400\,cm^{-1}$), C–H (peak at $3900\,cm^{-1}$), and a ketone (peak at $1710\,cm^{-1}$). It must be a primary alcohol as it can be oxidised to an acid. The substance is $CH_3COCH_2CH_2OH$/4-hydroxybutan-2-one or $CH_2OHCOCH_2CH_3$/1-hydroxybutan-2-one.

Exam tip

The peak at $3400\,cm^{-1}$ is broad due to hydrogen bonding. This is often asked in exams.

Following the progress of a reaction

IR spectroscopy can be used to monitor the progress of a reaction — for example, the oxidation of an alcohol. Secondary alcohols are oxidised by acidified potassium dichromate(VI) to give ketones. This is readily seen in the IR spectrum of the reactant alcohol, which has a broad O–H stretch at 3200–$3800\,cm^{-1}$ but no carbonyl absorption around $1700\,cm^{-1}$. As the reaction progresses, the peak due to the O–H stretch disappears and a new peak appears at around $1700\,cm^{-1}$, due to the carbonyl group in the product.

Summary

You must be able to:
- analyse a mass spectrum and identify the ions responsible for peaks at specified m/z values
- evaluate the m/z of the molecular ion and hence calculate the molecular formula from the empirical formula
- analyse an infrared spectrum and identify the bonds responsible for the peaks (especially C=O and O–H)
- use an infrared spectrum to identify a compound given its molecular formula

Topic 8 Energetics I

Enthalpy change

The heat change ($\Delta_r H$) that occurs during a chemical reaction at constant pressure (i.e. in a vessel open to the atmosphere) is called the **enthalpy change** for that reaction. It is measured in $kJ\,mol^{-1}$, where 'mol^{-1}' refers to the molar quantities given in the equation. Therefore:

$$2SO_2(g) + O_2(g) \rightarrow 2SO_3(g) \quad \Delta_r H = -92 \text{ kJ}\,mol^{-1}$$

means that 92 kJ of heat is given out when 2 moles of sulfur dioxide combine with 1 mole of oxygen to give 2 moles of sulfur trioxide. Such an equation is called a **thermochemical** equation.

Note that an enthalpy change should not be called an energy change. Energy changes during reactions are calculated at constant volume and the values obtained are not necessarily the same as the enthalpy change for the reaction.

Changes of state (e.g. from liquid to gas or from solid to liquid) are accompanied by heat changes. The states of the substances in the reaction are important. For example, the combustion of methane would give out more energy if the product (water) were liquid rather than gaseous, the difference being the enthalpy of vaporisation of the liquid water. Thermochemical equations are commonly shown for substances in their **standard state**, the associated enthalpy changes being standard enthalpies. This is shown by the superscript symbol '$\ominus$', i.e. $\Delta H^{\ominus}$.

Standard conditions mean:

- all reagents and products are in their thermodynamically most stable state
- 100 kPa (1 atmosphere) pressure
- a specified temperature — usually 25°C (298 K)

Some elements have different forms (allotropes) in the solid state. For carbon it can be necessary to specify graphite or diamond rather than just solid (s).

Enthalpy level diagrams

Enthalpy level diagrams show the relative energy levels of reactants and products. The horizontal axis is often absent, unlabelled, or may be called the 'reaction coordinate'. Enthalpy level diagrams for exothermic and endothermic reactions are shown in Figure 38.

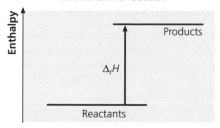

Figure 38

> **Exam tip**
>
> Exothermic reactions have a negative value of ΔH.

> **Exam tip**
>
> Always include state symbols in thermodynamic equations.

> **Exam tip**
>
> If a definition of a *standard* enthalpy change is asked, the details of standard conditions must be given in the answer.

> **Exam tip**
>
> Make sure that the arrow for ΔH goes from the enthalpy level of the reactants to the products.

Heat may be given out or be taken in during a chemical reaction. The international convention is that heat changes are seen from the point of view of the chemical system — imagine you are sitting inside the reaction mixture.

- Heat taken in is regarded as positive; so endothermic processes have a positive ΔH.
- Heat given out is regarded as negative; so exothermic processes have a negative ΔH.

Note that the positive or negative signs represent *conventions* concerning the direction of heat flow. They do not represent relative magnitudes in the same way that plus or minus signs do with numbers. Suppose that for one reaction $\Delta H = -100 \, kJ \, mol^{-1}$ and that for another it is $-200 \, kJ \, mol^{-1}$. The second value is not smaller than the first — indeed it represents twice as much heat given out. Instead of using the terms 'smaller' or 'larger', which are ambiguous, use descriptive terms, such as 'more exothermic' or 'more negative'.

> **Exam tip**
>
> You should always include the sign when you write a value for ΔH, even if it is positive.

Definitions of standard enthalpy changes

- **Standard enthalpy of reaction, $\Delta_r H^{\ominus}$**, is the heat change when the molar quantities (as in the equation written) react completely at 100 kPa pressure and a specified temperature (usually 298 K).

    For the equation $Fe(s) + 1\frac{1}{2}Cl_2(g) \rightarrow FeCl_3(s)$, it is the heat change when 1 mol of iron reacts with 1½ mol of chlorine.

    If the equation is written $2Fe(s) + 3Cl_2(g) \rightarrow 2FeCl_3(s)$, it is the heat change when 2 mol of iron reacts with 3 mol of chlorine.

- **Standard enthalpy of formation, $\Delta_f H^{\ominus}$**, is the heat change for the formation of 1 mole of substance from its *elements*, all substances being in their most stable state at 100 kPa pressure and a specified temperature.

- **Standard enthalpy of combustion, $\Delta_c H^{\ominus}$**, is the heat change when 1 mole of substance is completely burnt in *excess oxygen*, all substances being in their most stable state at 100 kPa pressure and a specified temperature.

- **Standard enthalpy of neutralisation, $\Delta_{neut} H^{\ominus}$**, is the heat change when an acid is neutralised by an alkali, to produce *1 mole of water* at 100 kPa pressure and a specified temperature.

- **Standard enthalpy of atomisation, $\Delta_{at} H^{\ominus}$**, is the heat change for the formation of 1 mole of gaseous atoms from an element in its standard state at 100 kPa pressure and a specified temperature.

    Note that the definition of $\Delta_{at} H^{\ominus}$ refers to the *formation of 1 mole of gaseous atoms*. Therefore, in the case of chlorine, the heat change measured is for the process:

$$\frac{1}{2}Cl_2(g) \rightarrow Cl(g)$$

> **Exam tip**
>
> Definitions can be given as a statement plus an equation, for example the standard enthalpy of formation of sodium chloride would be: 'The enthalpy change per mole at 100 kPa pressure and a stated temperature for $Na(s) + \frac{1}{2}Cl_2(g) \rightarrow NaCl(s)$'.

> **Exam tip**
>
> Note that all the definitions refer to 1 mol.

Some simple thermochemistry experiments

There are several simple experiments that can be used to evaluate enthalpy changes. The principle is always to mix known amounts of the reactants and measure the change in temperature. The heat transferred from the reaction vessel to the surroundings (an exothermic reaction) or from the surroundings to the reaction mixture (an endothermic reaction) is calculated as:

heat energy transferred = mass × specific heat capacity × temperature change

$Q = mc\Delta T$

The following account is not intended to substitute for worksheets or other practical instructions. For further details you should consult a book of practical chemistry.

The principles for evaluating enthalpy change are as follows:

- Use of a known amount of substance.
- Insulation against heat losses. Since the temperature rise or fall for most reactions in solution is fairly small, heat losses are much less than is commonly imagined. The same is not true for combustion reactions (e.g. the burning of alcohols to find the heat of combustion), where the heat losses are significant.
- Correction of the maximum observed temperature change to compensate for heat loss for slow reactions. This is particularly important for reactions between solids and liquids (e.g. heat of displacement in the reaction between copper(II) sulfate solution and zinc metal).
- Calculation of the results, using heat transferred = $mc\Delta T$, where m = mass of the liquid that heated up, c = its heat capacity and ΔT the temperature change.
- Calculation of the molar heat change. $\Delta H = \pm$ heat transferred/moles of reactant with a − sign if the temperature rose and a + sign if it fell.

Determination of the enthalpy of neutralisation of an acid

- A known volume of acid of known concentration is pipetted into a polystyrene cup and its temperature measured.
- A known volume of alkali, containing a slight excess, is placed in a beaker and its temperature measured.
- The alkali is poured into the acid and thoroughly stirred.
- The maximum temperature is read.

Calculation

The enthalpy of the reaction is found using the following steps:

- The temperature rise, ΔT, = final temperature − the mean temperature of the acid and alkali
- The mass of solution that warmed up is assumed to be equal to the volume of acid solution + volume of alkali solution, in cm^3.
- The heat transferred, Q, = mass of solution × its specific heat capacity × temperature rise
- The amount of acid reacted = volume in dm^3 × concentration in $mol\,dm^{-3}$.
- The heat change per mole of water formed is then found by dividing Q by the amount (number of moles) of hydrogen ions used and including a sign and units.

Exam tip

The mass, m, is assumed to be the mass of the solvent or the mass of water being heated in a combustion experiment. It is a common error to substitute the mass of the reactants into this equation.

Exam tip

Always give the sign as well as the numerical value of ΔH: + if the temperature falls and − if it rises.

Exam tip

This technique can be used for any type of reaction that can be carried out in an expanded polystyrene cup, including neutralisation reactions and displacement reactions.

Exam tip

The specific heat capacity of water is given in the data booklet.

Content Guidance

Example

When $25.0\,\text{cm}^3$ of a $1.00\,\text{mol}\,\text{dm}^{-3}$ solution of HCl was added to $25.0\,\text{cm}^3$ of $1.10\,\text{mol}\,\text{dm}^{-3}$ NaOH the temperature rose by $6.8°C$. Calculate the enthalpy change for the neutralisation of hydrochloric acid. (The heat capacity of the solution is $4.18\,\text{J}\,°C^{-1}\,g^{-1}$.)

Answer

$$\text{heat produced} = mc\Delta T = 50.0\,\text{g} \times 4.18\,\text{J}\,°C^{-1}\,g^{-1} \times 6.8°C = 1421.2\,\text{J}$$

$$\text{amount of HCl} = \text{concentration} \times \text{volume in dm}^3$$
$$= 1.00\,\text{mol}\,\text{dm}^{-3} \times 0.0250\,\text{dm}^3$$
$$= 0.0250\,\text{mol}$$

$$\Delta_{\text{neut}}H \text{ of HCl} = \frac{-1421.2\,\text{J}}{0.0250\,\text{mol}}$$
$$= -56.848\,\text{J}\,\text{mol}^{-1}$$
$$= -56.8\,\text{kJ}\,\text{mol}^{-1}$$

> **Exam tip**
>
> Remember that ΔH is *negative* because the temperature rose during the experiment.

Determination of the enthalpy change of a slow reaction

An example would be the displacement reaction of zinc and copper sulfate solution.

- Some zinc powder is weighed out. This must be in excess.
- A known volume of a given concentration of copper sulfate solution is taken and placed in a polystyrene cup.
- The temperature of the copper sulfate solution is measured for 4 minutes.
- At the fifth minute, the zinc is added and the mixture stirred.
- The temperature is measured every 30 seconds until the solution becomes colourless.
- A graph of temperature against time is plotted, and the lines are extrapolated to enable the temperature change, ΔT, at minute 5 to be calculated, and corrected for any heat losses (Figure 39).

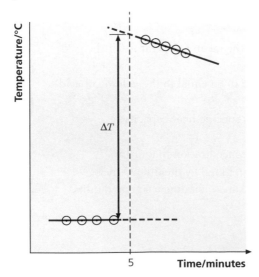

Figure 39

- The heat transferred, Q, is $mc\Delta T$, where m is the mass of the solution (usually taken to be equal to its volume for aqueous solutions), c is the heat capacity of the solution (usually taken to be equal to that of water for dilute aqueous solutions), and ΔT is the corrected temperature change.
- The molar heat change ΔH is then found by dividing the result by the amount (number of moles) of copper ions used and including a sign and units.

The method of measuring the temperature change is designed to compensate for two sources of error.

1 Loss of heat to the surroundings (or sometimes gain, since some processes arc cndothcrmic and the temperature falls below room temperature).

2 Slowness of reaction as this gives more time for heat losses.

Enthalpy of combustion of alkanes and alcohols

This simple experiment is not accurate, since the loss of heat to the surroundings is considerable.

- A spirit lamp is used, filled with either a liquid alkane or alcohol.
- A known volume of water is added to a copper calorimeter can.
- The spirit lamp is weighed and placed under the can (without a gauze).
- The temperature of the water is measured and the lamp is lit.
- When the temperature of the water has risen by a suitable amount, the lamp is extinguished and re-weighed.

The heat gained by the water is equal to $mc\Delta T$, where m is the mass of water in the calorimeter can, c is the heat capacity of water and ΔT is the temperature change. If the mass of the liquid burnt is found from the weighings, then the heat evolved per mole of fuel can be found.

Knowledge check 18

Ethanol, C_2H_5OH, of mass 0.23 g was burnt in a spirit lamp and heated 100 cm^3 of water by 15 K. Given that the heat capacity of the water is 4.2 J K^{-1} g^{-1}, calculate the enthalpy of combustion of ethanol.

The errors in this experiment are large, arising mainly from the following:

- Heat losses — heat is lost by convection around the calorimeter, so this energy never finds its way into the water.
- Incomplete combustion of the fuel — spirit lamps do not have a good oxygen supply, and the flame, particularly from alkanes, is often sooty.
- Evaporation — some of the fuel evaporates from the spirit lamp, so the mass burnt cannot be determined accurately.
- The water produced in the combustion is a gas, whereas in the equation for the standard enthalpy of combustion it is a liquid. This results in a less exothermic value of Δ_cH.

Exam tip

Remember that the mass is that of the solution, not that of the zinc.

Exam tip

Remember that the heat transferred Q is in joules, J, whereas the usual unit for ΔH is in kJ mol^{-1}. This is why it is essential to give a unit with your answer.

Exam tip

If a thermometer accurate to ±0.5°C is used, the error in ΔT is 2×0.5.

Knowledge check 17

In a thermochemical experiment, a thermometer accurate to 0.5°C was used. The initial temperature was 18.5°C and the final temperature was 26.0°C. Calculate the % error in ΔT.

Exam tip

Remember that all combustion reactions are exothermic, so ΔH will be negative.

Hess's law

Hess's law states that the enthalpy change for the process $Y \rightarrow Z$ is independent of the route used to effect the change, provided that the states of Y and Z are the same for each route.

The use of **standard enthalpies** avoids the problem of the states at the beginning and end of the experiment, since the states of Y and Z are defined.

This means that enthalpy changes can be found using other data. For example, enthalpies of reaction can be found from enthalpies of formation or enthalpies of combustion. It is possible, using Hess's law, to find the enthalpy change for reactions that cannot be performed directly.

Note that Hess's law is a particular case of the first law of thermodynamics, which states that energy can be neither created nor destroyed.

Using enthalpies of formation

The Hess's law diagram for calculating enthalpy of reaction from enthalpies of formation is shown in Figure 40.

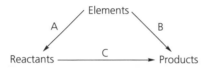

Figure 40

Hess's law says that $C = -A + B$, where A, B and C are the enthalpy changes. Using the definition of enthalpies of formation:

A = (sum of the enthalpies of formation of the reactants)

B = (sum of the enthalpies of formation of the products)

Therefore:

C = − (sum of the enthalpies of formation of the reactants) + (sum of the enthalpies of formation of the products)

Consider the combustion of ethanol:

$$CH_3CH_2OH(l) + 3O_2(g) \rightarrow 2CO_2(g) + 3H_2O(l)$$

The example shows how the heat change for the combustion of ethanol can be found (Figure 41). The states of all substances must be shown, because changes of state also involve enthalpy changes.

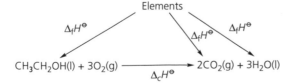

Figure 41

Exam tip

When, and only when, $\Delta_f H$ data are used, you may quote the expression:

$\Delta_r H$ = −sum of $\Delta_f H$ of reactants + sum of $\Delta_f H$ of products

$$\Delta_cH^\ominus = -\Delta_fH^\ominus \text{ (CH}_3\text{CH}_2\text{OH)} + 2\Delta_fH^\ominus \text{ (CO}_2\text{)} + 3\Delta_fH^\ominus \text{ (H}_2\text{O)}$$

Inserting the appropriate values gives:

$$\Delta_cH^\ominus = -(-277.1) + 2(-393.5) + 3(-285.8) = -1367.3 \text{ kJ mol}^{-1}$$

Note that $\Delta_fH^\ominus$ of oxygen is not included, as it is an element in its standard state and so its $\Delta_fH^\ominus$ = zero.

Using enthalpies of combustion

The Hess's law diagram for calculating enthalpy of reaction from enthalpies of combustion is shown in Figure 42.

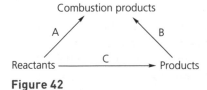

Combustion products

Figure 42

A = (sum of the enthalpies of combustion of the reactants)

B = (sum of the enthalpies of combustion of the products)

Therefore:

C = A − B = (sum of the enthalpies of combustion of the reactants) − (sum of the enthalpies of combustion of the products)

Consider the reaction between ethanol and ethanoic acid to give ethyl ethanoate and water:

$$\text{CH}_3\text{CH}_2\text{OH(l)} + \text{CH}_3\text{COOH(l)} \rightarrow \text{CH}_3\text{COOCH}_2\text{CH}_3\text{(l)} + \text{H}_2\text{O(l)}$$

The Hess's law diagram for this reaction using enthalpies of combustion is shown in Figure 43.

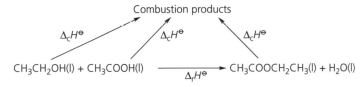

Combustion products

Figure 43

Exam tip

If you had used the expression 'Δ_fH of products – Δ_fH of reactants', you would have got the wrong sign in your answer.

Therefore:

$$\Delta_rH^\ominus = \Delta_cH^\ominus(\text{CH}_3\text{CH}_2\text{OH(l)}) + \Delta_cH^\ominus(\text{CH}_3\text{COOH(l)}) - \Delta_cH^\ominus(\text{CH}_3\text{COOCH}_2\text{CH}_3\text{(l)})$$

Water, of course, does not burn so should not be included when calculating enthalpies of combustion. Inserting the appropriate values gives:

$$\Delta_rH^\ominus = (-1367.3) + (-874.1) - (-2237.9) = -3.5 \text{ kJ mol}^{-1}$$

Determining enthalpy changes for reactions that cannot be performed directly

Core practical 8: To determine the enthalpy change of a reaction using Hess's law

Many Hess's law calculations relate to reactions that cannot be performed directly. For example, the direct determination of ΔH for the decomposition of sodium hydrogencarbonate on strong heating is not possible. The reaction is:

$$2NaHCO_3(s) \rightarrow Na_2CO_3(s) + CO_2(g) + H_2O(g)$$

An approximate method for finding ΔH is to react sodium hydrogencarbonate and sodium carbonate separately with hydrochloric acid, finding the heat change in each case. The two values obtained can be combined to give an approximate value for ΔH. It is not the actual value because the states of the various reagents are not exactly the same as in the thermal decomposition, but, if desired, adjustments can be made for this using other data, such as hydration enthalpies and enthalpies of solution.

A known amount of sodium hydrogencarbonate is reacted with an excess of hydrochloric acid in an expanded polystyrene cup and the temperature change is determined using methods described previously (page 46). A similar experiment using sodium carbonate is performed and ΔH for each reaction is calculated.

Equation 1: $NaHCO_3(aq) + HCl(aq) \rightarrow NaCl(aq) + CO_2(g) + H_2O(l)$ ΔH_1

Equation 2: $Na_2CO_3(aq) + 2HCl(aq) \rightarrow 2NaCl(aq) + CO_2(g) + H_2O(l)$ ΔH_2

The overall thermal decomposition reaction approximates to (2 × equation 1) − (equation 2), so that:

$$\Delta H = 2\Delta H_1 - \Delta H_2$$

Knowledge check 19

Use the enthalpies of solution of hydrated and anhydrous copper(II) sulfate

$CuSO_4.5H_2O(s) \rightarrow CuSO_4(aq)$ $\Delta H = +6\,kJ\,mol^{-1}$

$CuSO_4(s)$ $\rightarrow CuSO_4(aq)$ $\Delta H = -73\,kJ\,mol^{-1}$

to calculate the enthalpy change for:

$CuSO_4.5H_2O(s) \rightarrow CuSO_4(s) + 5H_2O(l)$

Using mean bond enthalpies

To calculate enthalpies of reaction

The **bond enthalpy** of a bond in a particular molecule is the heat energy required to break that bond. It is always a positive number.

The **mean bond enthalpy** is the enthalpy change when 1 mole of the specified type of bond is broken; an average value is taken which has been determined over a wide variety of molecules.

Hess's law can be used to find approximate values of $\Delta H^{\ominus}$ for a reaction by use of mean bond enthalpies. Using mean bond enthalpies is approximate because bond enthalpies may vary quite a lot. Thus the C=O bond enthalpy in CO_2 is 805 kJ mol⁻¹ but in methanal, HCHO, it is 695 kJ mol⁻¹. Breaking all the bonds in all the reactants leads to a collection of atoms. Thus polyatomic elements such as oxygen have to be included in the calculations, unlike in the case of enthalpies of formation.

The Hess's law diagram for the use of bond enthalpies is shown in Figure 44.

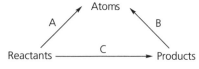

Figure 44

A = (sum of bond enthalpies of the reactants)

B = (sum of the bond enthalpies of the products)

Therefore:

C = (sum of bond enthalpies of the reactants) – (sum of the bond enthalpies of the products)

Bond enthalpies are always endothermic, i.e. positive.

When working out enthalpies of reaction from bond enthalpies, use the following rules:
- Write the reaction using structural formulae, so that you remember all the bonds.
- Ignore groups that are unchanged between the reactants and the products. In the addition reaction of ethene with hydrogen, there is no need to involve the CH_2 groups since they survive unchanged. There is no point in breaking all the bonds only to make them again.

Consider the combustion of ethanol (Figure 45).

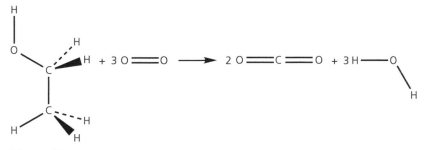

Figure 45

The bonds broken are five C–H, one C–C, one C–O (note the single bond) and one O–H. Those made are four C=O (note the double bond) and six O–H. Use the equation:

$$\Delta_r H = \text{bonds broken} - \text{bonds made}$$

and the bond enthalpies in Table 8 to calculate the approximate enthalpy of reaction (Table 9).

Knowledge check 20

Use the bond energies/ kJ mol⁻¹ given below to calculate the enthalpy change of the reaction:

$$H_2C=CHCH_3 + HCl \rightarrow CH_3CHClCH_3$$

C-H: 413, C=C: 612, C-C: 347, H-Cl: 431, C-Cl 346

Bond	C–H	C–C	C–O	O–H	C=O
Mean bond enthalpy/kJ mol⁻¹	413	347	358	464	805

Table 8

Bonds broken	Value/kJ mol⁻¹
5 × C–H	5 × 413 = 2065
1 × C–C	347
1 × C–O	358
1 × O–H	464
Total	3234

Bonds made	Value/kJ mol⁻¹
4 × C=O	4 × 805 = 3220
6 × O–H	6 × 464 = 2784
Total	6004

Table 9

Approximate ΔH of reaction = 3234 – 6004 = –2770 kJ mol⁻¹

> **Exam tip**
>
> The value for ΔH is only approximate as *average* bond enthalpies were used in the calculation.

Calculation of bond enthalpy from $\Delta_r H$

If all but one of the bond enthalpies for the substances in a reaction are known together with its enthalpy of reaction, the unknown bond enthalpy can be calculated.

Consider the reaction of buta-1,3-diene with hydrogen.

$$CH_2=CH–CH=CH_2(g) + 2H_2(g) \rightarrow CH_3–CH_2–CH_2–CH_3(g) \quad \Delta_r H = -238\,kJ\,mol^{-1}$$

Two C=C and two H–H bonds have to be broken followed by four C–H and two C–C bonds made.

Mean bond enthalpy data are shown in Table 10.

Bond	C–H	C–C	H–H
Mean bond enthalpy/kJ mol⁻¹	+413	+347	+436

Table 10

Let the C=C bond enthalpy in buta-1,3-diene = z.

$\Delta_r H$ = bonds broken – bond made

$-238 = 2z + 2 \times 436 - (4 \times 413 + 2 \times 347)$

$\qquad = 2z - 1474$

$2z = 1474 - 238$

$\quad = +1236$

$z = +618$

The C=C bond enthalpy in buta-1,3-diene is +618 kJ mol⁻¹.

Bond enthalpies and reaction mechanisms

Reaction of halogenoalkanes with hydroxide ions

The rate of reaction between the halogenoalkanes and hydroxide ions in aqueous ethanol depends on the bond strength of the carbon–halogen bond, all other things being equal. If X is a chlorine, bromine or iodine atom:

$$CH_3CH_2X + OH^- \rightarrow CH_3CH_2OH + X^-$$

The average bond enthalpy of C–Cl is $+346\,kJ\,mol^{-1}$, C–Br is $+290\,kJ\,mol^{-1}$, and C–I is $+228\,kJ\,mol^{-1}$.

When chloro-, bromo- and iodopropane are placed separately in solutions of silver nitrate in aqueous ethanol and left in a warm-water bath, the precipitate of silver iodide appears first, followed by that of silver bromide and then, after a much longer time, silver chloride. The stronger the carbon–halogen bond, the slower is the reaction, as the activation energy for the reaction increases in the order iodo- < bromo- < chloro-.

Exam tip

The rate of reaction is determined by the C–X bond energy and *not* by its polarity.

Photohalogenation of methane with chlorine

The overall equation for this reaction is:

$$CH_4(g) + Cl_2(g) \rightarrow CH_3Cl(g) + HCl(g)$$

The initiation step is:

$$Cl_2 \rightarrow 2Cl\bullet$$

This is followed by one of two possible propagation steps:

Possible step 1: $CH_4 + Cl\bullet \rightarrow \bullet CH_3 + HCl$

Possible step 2: $CH_4 + Cl\bullet \rightarrow CH_3Cl + H\bullet$

We can use bond enthalpies to predict which propagation step is the more likely:

- Possible step 1: a C–H bond is broken ($+435\,kJ\,mol^{-1}$) and a H–Cl bond made ($-432\,kJ\,mol^{-1}$), giving $\Delta H = +3\,kJ\,mol^{-1}$
- Possible step 2: a C–H bond is broken ($+435\,kJ\,mol^{-1}$) and a C–Cl bond is made ($-346\,kJ\,mol^{-1}$), giving $\Delta H = +89\,kJ\,mol^{-1}$

Step 1, being less endothermic, is the more likely. This is supported by the observed products — step 2 would give rise to some hydrogen as a result of two $H\bullet$ radicals combining, but none is seen, whereas step 1 would lead to some C_2H_6 forming as a result of two $\bullet CH_3$ radicals colliding.

Errors and assumptions in heat measurements

Errors in measurements

Table 11 shows possible sources of error when calculating enthalpy change.

Temperature measurements	Thermometers have an error of ±0.5°C per reading, which is ±1.0°C per measurement of ΔT
Volume measurements	Pipettes have an error of ±0.1 cm³ Measuring cylinders have an error of ±1 cm³ Beakers have a much larger error
Mass measurements	Mass measurements have an error of ±0.01 g per reading, which is ±0.02 g per mass of substance
Heat losses	Small for rapid reactions carried out in a polystyrene cup Significant for slow reactions carried out in a polystyrene cup, but can be compensated for by plotting a temperature/time graph (see page 46) Very large when using a spirit lamp

Table 11

Assumptions

- Mass of liquid in polystyrene cup or beaker = volume of liquid. This is a reasonable assumption as the density of solutions used is very close to $1\,g\,cm^{-3}$.
- Specific heat capacity of liquid in cup = specific heat capacity of water. This is an accurate assumption.
- In combustion experiments it is assumed that all the water produced is liquid whereas most is gaseous. This is a poor assumption.

Summary

After studying this topic, you should:
- know what the standard conditions are
- know that ΔH values for exothermic reactions are negative
- know that when the temperature of a reaction mixture rises, ΔH will be negative
- be able to draw enthalpy level diagrams of exothermic and endothermic reactions.
- be able to define the standard enthalpy of:
  - formation
  - combustion
  - neutralisation
  - atomisation
- be able to calculate ΔH of a reaction using:
  - experimental data
  - Hess's law
  - mean bond enthalpies
- be able to evaluate sources of error and assumptions made in enthalpy experiments

■ Topic 9 Kinetics I

The factors that control the rate of a chemical reaction include:

- the concentration of the reagents (for reactions in solution)
- the temperature of the reaction system
- the pressure (for reactions in the gas phase)
- the surface area of any solid reagents
- the presence of a catalyst

Collision theory

Chemicals cannot react unless they collide. The theory of reaction rates is therefore called **collision theory**. When thinking about reaction rates, you should do so on a molecular level and try to imagine the collisions occurring.

The important factors are:

- the number of collisions per unit time — the **collision frequency**
- the energy with which the particles collide — the **collision energy**
- the **orientation** in which the particles collide, particularly important for large molecules

Collision frequency

Collision frequency increases with concentration in a liquid system, or with an increase in pressure in a gaseous one. In each case, the distance between colliding species is reduced, so there is less distance to travel before encountering another molecule. This is a significant factor.

Collision frequency increases with surface area for a solid reagent. In this case, the increased area raises the probability of a molecule in the other phase (gas or liquid) colliding with the solid. This is a significant factor.

Collision frequency increases with temperature. The molecules are moving faster and so travel the necessary distance more quickly. This has a marginal effect – see below.

Collision energy

The minimum collision energy needed for particles to react is called the **activation energy (E_a)**. Particles that collide with an energy equal to or greater than E_a react if their orientation is correct.

Increasing the temperature increases the proportion of particles that collide with energies greater than E_a. The collision frequency rises roughly linearly with rising temperature, but the increase in the number of particles with collision energies above E_a is exponential. As a rough guide, when the temperature of a reaction that proceeds steadily is increased by 10°C, the rate of collision increases by about 1% whereas the increase in the number of particles with sufficient activation energy increases by about 100%.

> **Exam tip**
>
> Only a very small proportion of the collisions that occur in a reaction system are successful, i.e. lead to the formation of products.

> **Exam tip**
>
> The effect of an increase in temperature on collision *energy* is more important than the effect on collision *frequency*.

Orientation at collision site

Particles must collide in such a way that their reactive parts come into contact. This is why 1-bromobutane reacts with OH^- ions at a slower rate than bromoethane and OH^- ions under the same conditions.

Rate of reaction

Definitions

- The rate of reaction is the amount by which the concentration of a reactant or product changes in a given time.

$$\text{rate} = \frac{\text{change in concentration}}{\text{time taken}}$$

Its units are $mol\,dm^{-3}\,s^{-1}$.

- The initial rate is the rate of reaction at the time when the reactants are mixed and is the gradient at time $t = 0$ of a graph of concentration against time.
- The rate at time t is the gradient at time t of a graph of concentration against time.
- A good approximation of initial rate is the amount by which the concentration of a reactant or product has changed in a given time from mixing, *provided* the concentration has not changed by more than 10% in that time.

Calculation of rate

1 Obtain data of the time taken for a certain amount of one of the reactants to be used up or a certain amount of a product to be produced.

2 Follow the reaction and obtain data of how the concentration varies with time. Plot a graph of the concentration of a reactant or product against time. Draw a tangent at a given point on the graph — the gradient equals the rate of reaction at that point.

Time for the reaction to finish

An approximate method for calculating rate is to add the two reactants and time how long it takes for the reaction to stop. An example would be to add a strip of magnesium to an *excess* of dilute sulfuric acid and time how long it takes for the production of hydrogen bubbles to stop. The experiment is repeated with either a different concentration of the acid or at a different temperature. The assumption is then made that the rate is proportional to 1/time. However, this is only valid if the concentration of the acid has fallen by less than 10–15% and if the temperature did not change by more than 5°C during the measurement.

The iodine 'clock'

In a 'clock' reaction, the reactants are mixed and the time taken to produce a fixed amount of product is measured. The experiment is then repeated several times using different starting concentrations.

This gives several initial rates of reaction at different concentration.

The oxidation of iodide ions by hydrogen peroxide in acid solution can be followed as a 'clock' reaction:

$$H_2O_2(aq) + 2I^-(aq) + 2H^+(aq) \rightarrow I_2(s) + 2H_2O(l)$$

- $25\,cm^3$ of hydrogen peroxide solution is mixed in a beaker with $25\,cm^3$ of water and a few drops of starch solution are added.
- $25\,cm^3$ of potassium iodide solution and $5\,cm^3$ of a dilute solution of sodium thiosulfate are placed in a second beaker.
- The contents of the two beakers are mixed and the time taken for the solution to go blue is measured.
- The experiment is repeated with the same volumes of potassium iodide and sodium thiosulfate but with $20\,cm^3$ of hydrogen peroxide and $30\,cm^3$ of water, and then with other relative amounts of hydrogen peroxide and water, totalling $50\,cm^3$.

The reaction produces iodine, which reacts with the sodium thiosulfate. When all of the sodium thiosulfate has been used up, the next iodine that is produced reacts with the starch to give an intense blue-black colour.

The amount of iodine produced in the measured time is proportional to the volume of sodium thiosulfate solution taken. Therefore, the average rate of reaction for each experiment is proportional to 1/time.

Note that if the moles of thiosulfate ions is much less than the moles of the hydrogen peroxide and iodide ions, the average rate measured is almost identical to the initial rate.

The sulfur 'clock'

Sodium thiosulfate is decomposed by acid, producing a precipitate of sulfur:

$$S_2O_3{}^{2-}(aq) + 2H^+(aq) \rightarrow S(s) + SO_2(aq) + H_2O(l)$$

This reaction can be followed as a 'clock' reaction.
- A large X is drawn on a white tile with a marker pen.
- $2\,cm^3$ of sodium thiosulfate solution is mixed with $25\,cm^3$ of water in a beaker.
- $25\,cm^3$ of dilute nitric acid is placed in a second beaker.
- The first beaker is placed on top of the X and the contents of the second one are added.
- The mixture is stirred and the time taken for sufficient sulfur to be produced to hide the X when looking down through the beaker is measured.
- The experiment is repeated with different relative amounts of sodium thiosulfate and water, totalling $50\,cm^3$.

The number of moles of sulfur produced is the same in all experiments. Therefore, the average rate of reaction for each experiment is proportional to 1/time.

Concentration–time graphs

The concentration of a reactant is measured over a period of time and a graph is plotted with concentration on the y-axis and time on the x-axis (Figure 46). The gradient at any point on this graph is the rate at that point.

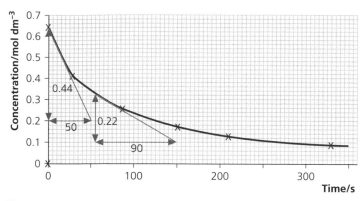

Figure 46

The gradient at time 0 = (0.20 − 0.64)/50 = −0.0088 mol dm^{-3} s^{-1}

The initial rate = − the gradient at time 0 = 0.0088 mol dm^{-3} s^{-1}

The gradient at time 50 s = (0.10 − 0.32)/90 = −0.0024 mol dm^{-3} s^{-1}

The rate at time 50 s = −the gradient at time 50 s = 0.0024 mol dm^{-3} s^{-1}

A comparison of rates at different initial concentrations or different temperatures can be obtained by plotting anything that is proportional to concentration against time. For example the volume of gas produced is proportional to the moles of gas produced.

Maxwell–Boltzmann distribution of molecular energies

Figure 47 shows the distribution of energy between molecules in a gas at temperature T_1 and at a higher temperature T_2.

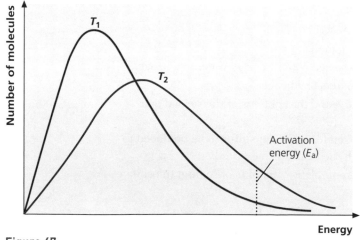

Figure 47

> **Exam tip**
>
> The gradient is negative because the concentration of the reactant decreases over time. If a graph of concentration of product against time had been plotted, the gradient would be positive and equal to the rate of reaction.

> **Exam tip**
>
> Do not use coloured pens in an exam — the colours will all look the same when photocopied.

> **Exam tip**
>
> Make sure that the graph approaches, but does *not* cut or touch the *x*-axis, at high energies.

There are several points to note about these graphs:

■ Both graphs start at the origin.
■ The distribution is skewed.
■ The higher temperature graph has its peak at higher energy, but the peak is lower than the graph for T_1. This is because the area under the two graphs must be the same, since this area is proportional to the total number of molecules.

Activation energy is shown in Figure 47. This is explained below.

Activation energy

The **activation energy**, E_a, is the minimum collision energy needed for particles to react on collision. Particles that collide with energy greater than or equal to E_a react if their orientation is correct.

Activation energy is represented above on the Maxwell–Boltzmann distribution. The area to the right of E_a represents the number of molecules that possess the activation energy or more and that therefore could react.

Effect of temperature

At a higher temperature, the area to the right of E_a increases, causing the proportion of successful collisions to increase. Therefore, the rate increases. Note that E_a is well to the right of the peak.

Catalysts

A **catalyst** changes the mechanism (or route) of a reaction to one having a lower value of E_a (Figure 48). This means that the proportion of successful collisions at a given temperature increases and therefore the rate increases.

Do not say that 'a catalyst lowers the activation energy' because this is not true and gets no credit. You need to state that the *mechanism* changes to one with a lower activation energy.

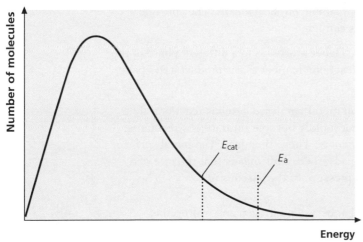

Figure 48

At least one of the reactants must combine with the catalyst as an initial step. The complex formed then reacts with another reactant to give the products and regenerate the catalyst.

Reaction profile diagrams

A reaction profile is a diagram with enthalpy increasing vertically. It shows the enthalpy level of the reactants and the products. A curved line is drawn to show the reaction pathway.

The reaction profile for a catalysed reaction therefore has at least two humps, with an intermediate reactant–catalyst complex. These are shown in Figure 49.

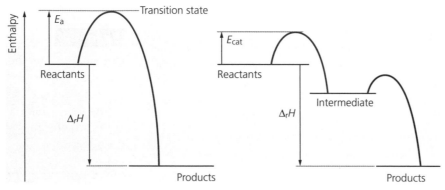

Figure 49

Exam tip

Note that the value for ΔH is the same for the catalysed and uncatalysed reaction, but that the activation energy for the catalysed reaction is much less.

Industrial processes

Heterogeneous catalysts

These are catalysts that are in a different phase from the reactants. Many industrial processes use this type of catalyst. They have the advantage that the catalyst can be separated from the equilibrium mixture by simple physical means. They provide a surface on which the reactants molecules can bond.

For example, the solid iron catalyst in the Haber process is in a different phase to the nitrogen and hydrogen, and stays in the reaction chamber as the product gases pass through.

A similar mechanism occurs in almost all metal-catalysed gaseous reactions. The catalyst has active sites on its surface that rapidly become saturated by reactants, which are then slowly converted into products. These then leave the metal surface, thus allowing more of the reactant to be adsorbed. This means that the rate of a reaction is not altered by an increase in pressure of the gaseous reactants. This is true of:

- the Haber process, in which the reaction between nitrogen and hydrogen is catalysed by iron
- the manufacture of hydrogen from steam and methane, which is catalysed by nickel
- the oxidation of ammonia by air, which is catalysed by platinum

A typical route would be:

gaseous reactants + surface of catalyst $\rightleftharpoons$ adsorbed reactants   *fast*

adsorbed reactants $\rightarrow$ adsorbed products   *slow*

adsorbed products $\rightarrow$ gaseous products + free catalyst surface   *fast*

Economic benefits

A catalyst allows an industrial process to be carried out at a lower temperature. This has a number of cost benefits.

1   Less energy is needed to maintain the temperature of the reaction as there will be less heat loss to the surroundings.

2   As most industrial processes are exothermic (the Haber process and the oxidation of ammonia for example), a lower temperature means that the position of equilibrium will lie more to the right, thus increasing the equilibrium yield.

Summary

After studying this topic, you should be able to:
- explain why an increase in pressure for gases or concentration for liquids increases the rate of a reaction
- calculate rates of reaction from time in 'clock' reactions and from concentration–time graphs
- define activation energy and catalyst
- draw Maxwell–Boltzmann distributions for reactions at different temperatures and with and without catalyst and use them to explain the change in rate of reaction

- draw reaction profile diagrams and use them to explain the effect of a catalyst on rate
- understand how solid catalysts can act as catalysts in industrial processes and their economic benefits

■ Topic 10 Equilibrium I

Dynamic equilibrium

A reaction in dynamic equilibrium:

- is undergoing no *net* change, so there is *no* change in the concentrations of the reactants and products
- is one in which the forward reaction (conventionally written from left to right) is happening at the same rate as the backward (reverse) reaction

A homogeneous equilibrium is one in which all the reagents and products are in the same phase. A phase has a clear boundary and it is not the same as the state of a material. A mixture of oil and water is entirely in the liquid state, but because there are two layers it is a two-phase system.

An example of a liquid-phase equilibrium system is that involving ethanol, ethanoic acid, ethyl ethanoate and water:

$$CH_3CH_2OH + CH_3COOH \rightleftharpoons CH_3COOCH_2CH_3 + H_2O$$

This is an esterification reaction.

An example of a gas-phase equilibrium system is the Haber process for the manufacture of ammonia:

$$N_2(g) + 3H_2(g) \rightleftharpoons 2NH_3(g)$$

Change of conditions

Effect of change in concentration

The esterification reaction is used as an example. If the concentration of either ethanol or ethanoic acid (or, in general, the substances on the left-hand side of the equation) is increased, the forward reaction occurs more rapidly until (new) equilibrium concentrations are established. The new equilibrium position will have a greater concentration of ethyl ethanoate and water (or, in general, the products — the substances on the right-hand side of the equation) than before. Thus, the yield is higher.

Reducing the concentration of the substances on the right-hand side, by removing them from the reaction vessel, has the same effect of increasing the yield.

Effect of change in pressure

The Haber process equilibrium is used as an example. Increasing the total equilibrium pressure moves the equilibrium composition towards the side with the smaller number of gaseous moles. In this case, the amount of ammonia would increase. A similar argument can be put forward for a decrease in pressure, which would move the equilibrium composition towards the reactants.

Effect of change in temperature

The effect of a change in the equilibrium temperature depends on whether the reaction, *defined in the forward direction* (left to right, as written), is exothermic or endothermic. The synthesis of ammonia is exothermic in the forward direction:

$$N_2(g) + 3H_2(g) \rightarrow 2NH_3(g) \qquad \Delta H = -92.4\,kJ\,mol^{-1}$$

> **Exam tip**
>
> Increasing the *concentration* of a substance on the *left* of the equation or *removing* a substance on the *right*, drives the equilibrium to the right.

> **Knowledge check 21**
>
> State and explain what is observed when concentrated HCl is added to the equilibrium:
>
> $CuCl_2 + 2Cl^- \rightleftharpoons CuCl_4^{2-}$
> blue   yellow

> **Knowledge check 22**
>
> State what you would observe when the pressure on the equilibrium below is suddenly increased:
>
> $N_2O_4(g) \quad \rightleftharpoons \quad 2NO_2(g)$
> colourless   brown

An increase in the equilibrium temperature moves the equilibrium position in the endothermic direction, in this example to the left, so the yield of ammonia falls at higher temperatures.

A similar argument can be made for endothermic reactions, where an increase in the (equilibrium) temperature moves the equilibrium position to the right. For example, the equilibrium between dinitrogen tetroxide (N_2O_4, colourless) and nitrogen dioxide (NO_2, brown) is endothermic from left to right, so increasing the temperature of the system makes the gas mixture a darker brown.

Remember that the temperatures are equilibrium temperatures, which are externally imposed, and therefore it is assumed that any change in equilibrium composition is not going to be able to affect that temperature.

Avoid Le Châtelier

You are likely to have come across Le Châtelier's principle, which states that changing the conditions under which an equilibrium is set up 'changes the composition in such a way as to tend to oppose the change'. Much confusion would be avoided if Le Châtelier were consigned to history. You will discover in the second year of A-level that it is an unnecessary idea in any case. It causes confusion because of the following:

- It leads some students to write about equilibria in a way that suggests that the reaction mixture can think — 'the equilibrium tries to reduce the temperature…'. This approach is poor and it matters because it obscures the true behaviour of equilibrium systems, which reach equilibrium because of the thermodynamics of the system. The composition of the system at equilibrium is that which minimises the energy of the system.

- It leads some students to believe that when the equilibrium composition changes, for example by raising the temperature, the system brings the temperature back down again to what it was. If it did do this, the composition would still be what it was initially. The temperature, the composition and the value of ΔH for the reaction are inextricably intertwined. The two different temperatures are *externally imposed* and represent two different *equilibrium* states with different compositions.

Industrial processes

Many industrial reactions are exothermic and are equilibrium reactions. The conditions used are a compromise between speed, yield and cost.

The Haber process

$$N_2(g) + 3H_2(g) \rightleftharpoons 2NH_3(g) \qquad \Delta H = -92.4\,kJ\,mol^{-1}$$

Even in the presence of a catalyst this reaction is so slow at room temperature that it is not observed, so a higher temperature must be used. Because the reaction is exothermic, the higher the temperature the lower is the yield. Thus a *compromise* temperature of 400 to 450°C is used whereby a reasonable yield is made quickly.

The yield is also increased by increasing the pressure, as there are fewer gas moles on the right than on the left. Because at 400°C the yield is very small, a pressure of 200 atm ($2 \times 10^4\,kPa$) is necessary in spite of the high cost of powering the compressors.

> **Exam tip**
>
> Exothermic reactions are driven to the left by an increase in temperature.
>
> Endothermic reactions are driven to the right by an increase in temperature.
>
> You need to state clearly which direction you are talking about in answers to examination questions.

> **Exam tip**
>
> Note that addition of a catalyst has *no* effect on the position of equilibrium.

The Contact process

$$2SO_2(g) + O_2(g) \rightleftharpoons 2SO_3(g) \qquad \Delta H = -196 \, kJ \, mol^{-1}$$

A catalyst is used so that the activation energy is lower, but a temperature of 425°C is still needed. However, the yield is high enough at this temperature for this process to be economical without the need for high pressure.

The equilibrium constant K

The equilibrium constant is a thermodynamic quantity related to the change in free energy of the system. This will be discussed in year 2 of the A-level course. At AS, the equilibrium constant in terms of concentrations, K_c, is defined according to the equilibrium equation.

Homogeneous systems

A homogeneous system is where all the components are in the same phase. Gaseous mixtures and solutions are both homogeneous.

For a homogeneous reaction in solution or in the gas phase:

$$nA + mB \rightleftharpoons qC + rD$$

where n, m, q and r are the numbers in the equation.

$$K_c = \frac{[C]^q \times [D]^r}{[A]^n \times [B]^m}$$

where $[C]$ is the *equilibrium* concentration of C in $mol \, dm^{-3}$, and so on.

Example 1

The equation for the reaction in the Haber process is:

$$N_2(g) + 3H_2(g) \rightleftharpoons 2NH_3(g)$$

The value of the equilibrium constant is:

$$K_c = \frac{[NH_3]^2}{[N_2][H_2]^3}$$

Example 2

Sulfur dioxide reacts with oxygen:

Equation 1: $\qquad 2SO_2(g) + O_2(g) \rightleftharpoons 2SO_3(g)$

$$K_c = \frac{[SO_3]^2}{[SO_2]^2[O_2]}$$

However, equation 1 can be divided throughout by two giving:

Equation 2: $SO_2(g) + \frac{1}{2}O_2(g) \rightleftharpoons SO_3(g)$

$$K_c = \frac{[SO_3]}{[SO_2][O_2]^{\frac{1}{2}}}$$

K_c for equation 2 = $\sqrt{K_c}$ for equation 1

Note that the expression for K_c depends on the equation used.

Heterogeneous systems

The reaction of iron with steam is an example of a heterogeneous equilibrium as the substances involved are in two different phases (gas and solid).

$3Fe(s) + 4H_2O(g) \rightleftharpoons Fe_3O_4(s) + 4H_2(g)$

As the concentration of a solid is a constant, it is omitted from the expression for K_c:

$$K_c = \frac{[H_2]^4}{[H_2O]^4}$$

Knowledge check 23

Write the expression for K_c for the reaction:

$CH_4(g) + H_2O(g) \rightleftharpoons CO(g) + 3H_2(g)$

....

Knowledge check 24

Write the expression for K_c for the partial dissolving of silver sulfate:

$Ag_2SO_4(s) \rightleftharpoons 2Ag^+(aq) + SO_4^{2-}(aq)$

....

Summary

After studying this topic, you should be able to:
- define dynamic equilibrium
- explain the effects, if any, of a change of concentration, pressure, temperature and the addition of catalyst to a given system in equilibrium
- discuss the conditions for a given industrial process in terms of rate, yield and cost
- deduce the expression for K_c for homogeneous and heterogeneous systems given the equation

Practical aspects

Hazard and risk

Safety is important in chemistry for obvious reasons. Chemistry deals with compounds that may be hazardous, so it is important that chemists are trained in the safe use of materials. **Hazard** is not the same as **risk**. A simple example — a deep lake with treacherous currents is a hazard, but it poses no risk unless you happen to be on it or in it. Hazard is an intrinsic property, risk is personal.

In any question on safety, you will be expected to suggest safety precautions specific to an experiment, and to be able to assess which is the most significant of various hazards presented.

In this context, the use of lab coats, safety glasses and pipette fillers is considered routine good practice and does not receive credit in examination answers.

The following is presented so as to give you the idea of the basis on which risk assessments might be made. *Safety will not be examined in this detail.* However, comments made by students on safety matters tend to be either so trivial as to be not worth reading, or so apocalyptic as to suggest that all chemistry should cease forthwith.

Every experiment should be the subject of a risk assessment by the institution that intends to perform it.

Hazard evaluation

The **hazard** associated with a substance is its potential to impair health. Some degree of hazard can be ascribed to almost any substance, while for some the toxicity or the harmful effects are not known fully.

Substances that are likely to be hazardous are those that are
- flammable — almost all organic substances
- carcinogenic materials — benzene
- toxic — barium salts, ethanedioic acid, halogens
- corrosive — concentrated sulfuric acid
- irritant — hydrogen chloride vapour

You may be asked to evaluate the risks in an experiment and state how you would minimise these risks.

Exposure potential

Hazardous substances vary enormously in potency. A substance with a high hazard may therefore present an acceptably small risk if the exposure potential is low. Conversely, unacceptable risks may result from high exposures to substances with low hazard.

Factors to be taken into account in evaluating exposure potential relate to both **substance** and **activity**.

Knowledge check 25

You are required to warm a boiling tube containing the volatile solvent ether, $C_2H_5OC_2H_5$. What is the main hazard in doing this and how could you minimise the risk?

Substance factors include:

- quantity used — small quantities reduce the risk
- volatility — very volatile organic substances increase the possibility of fire
- concentration if in solution — dilute acids are less corrosive than concentrated acids

Small-scale working is preferred wherever possible as this reduces the risk when using hazardous materials.

An alternative method using less hazardous materials should be used wherever possible.

Flammable substances

Flammable substances must be kept away from naked flames, so any heating necessary must be with a hot-water bath (rather than heated from underneath by a Bunsen burner) or an electric heating mantle.

Common hazards in school chemistry

The following list is by no means exhaustive or definitive.

Halogens are toxic and harmful if inhaled, although iodine is less toxic than chlorine or bromine, because it is a solid. Chlorine and bromine must always be used in a fume cupboard. Liquid bromine causes serious ulcerating burns and must be handled with gloves, so is best left to demonstration experiments by the teacher.

Ammonia is toxic. Concentrated ammonia solutions should be handled in the fume cupboard.

Concentrated mineral acids are corrosive. If spilt on the hands, washing with plenty of water — *never* alkali, which is even more damaging — is usually enough, but advice must be sought. Acid in the eye requires *immediate* and copious irrigation and immediate professional medical attention.

Barium chloride solution is extremely poisonous, as are **chromates** and **dichromates**.

Sodium hydroxide, potassium hydroxide or **concentrated ammonia** in the eye is *extremely serious*, and must always receive professional and *immediate* medical attention following copious irrigation of the eye. Sodium hydroxide and other alkali metal hydroxides are among the most damaging of all common substances to skin and other tissue. Treat them carefully.

None of the above is intended to prevent people doing chemistry — it is intended to encourage safe working practices.

In summary, risk can be minimised by:
- heating very flammable substances with a water bath or electric heating mantle
- wearing gloves when using corrosive substances
- carrying out experiments that involve toxic vapours in a fume cupboard
- working on a smaller scale
- carrying out the reaction using an alternative method that involves less hazardous substances

Knowledge check 26

You have been asked to investigate the reaction between concentrated sulfuric acid and solid sodium bromide, which produces bromine vapour. What are the hazards and how would you reduce the risks?

Practical techniques: organic chemistry

You are expected to be able to describe the techniques and draw suitable apparatus for:

- heating under reflux
- extraction with a solvent using a separating funnel
- distillation including distillation with addition of reactant
- boiling temperature measurement

Preparation of 1-bromobutane

1-bromobutane can be made from butan-1-ol and sodium bromide in 50% sulfuric acid. This preparation shows several practical techniques and the reason for them being used.

Step	Reason
(1) Dissolve some sodium bromide in water and mix with butan-1-ol in a round-bottomed flask. Fit a tap funnel to the flask via a still head. Place 25 cm³ of concentrated sulfuric acid in the tap funnel and allow the acid to fall drop by drop into the flask, keeping the contents well shaken and cooling occasionally in an ice-water bath.	**Why is the sulfuric acid added slowly? Why is cooling and shaking needed?** Sulfuric acid when diluted with water gives out a great deal of heat, enough sometimes to raise steam, which would cause dangerous splashing. Hot 50% sulfuric acid (produced in the flask) causes significant oxidation of the sodium bromide to bromine, which is useless in this experiment. The yield of 1-bromobutane could therefore be reduced.
(2) When the addition is complete, replace the tap funnel and still head with a reflux water condenser and gently boil the mixture over a sand bath, occasionally shaking the flask gently.	**Why is a sand bath used for heating?** The sand spreads the heating uniformly over the base of the flask. This reduces the likelihood of cracking and unwanted side reactions. **Why is the mixture heated?** Most organic reactions are slow because of the need to break strong covalent bonds — the activation energy for the reaction is high.
(3) Remove the reflux condenser and rearrange the apparatus for distillation. Distil off the crude 1-bromobutane.	**What impurities are present in the distillate?** The 1-bromobutane is contaminated with water, unchanged butan-1-ol and some sulfuric acid.
(4) Shake the distillate with water in a separating funnel and run off the lower layer of 1-bromobutane. Reject the aqueous layer.	**What does shaking with water achieve?** Water removes sulfuric acid and some of the butan-1-ol. **How do you decide which layer is to be kept?** 1-bromobutane is denser than water so it sinks to the bottom.
(5) Return the 1-bromobutane to the funnel, add about half its volume of concentrated hydrochloric acid and shake. Run off and discard the lower layer of acid.	**Why is concentrated hydrochloric acid added?** The acid protonates the butan-1-ol, giving an ionic species that is more soluble in water than the alcohol itself: $CH_3CH_2CH_2CH_2OH + H^+ \rightarrow CH_3CH_2CH_2CH_2OH_2^+$
(6) Shake the 1-bromobutane cautiously with dilute sodium carbonate solution, carefully releasing the pressure at intervals.	**Why is the mixture shaken with sodium carbonate solution?** This removes hydrochloric acid dissolved in the 1-bromo-butane: $Na_2CO_3 + 2HCl \rightarrow 2NaCl + CO_2 + H_2O$ **Why must the pressure be periodically released?** To avoid the stopper being pushed out and product being lost and sprayed all over you. The pressure is due to liberated carbon dioxide.
(7) Run off the lower layer of 1-bromobutane and add some granular anhydrous calcium chloride. Swirl the mixture until the liquid is clear.	**What is the function of the calcium chloride?** Calcium chloride is a drying agent.
(8) Filter the 1-bromobutane into a clean, dry flask and distil it, collecting the fraction boiling between 99 and 102°C.	**What is the significance of the temperatures quoted?** 1-bromobutane has a boiling temperature of 101.5°C, so the range is narrow enough to ensure that this is the distillate.

A similar reaction with sodium chloride can be used to make 1-chlorobutane. To make 1-iodobutane requires red phosphorus and iodine instead of sulfuric acid and sodium iodide (page 35).

Investigating rates of hydrolysis of halogenoalkanes

Take equal volumes of different halogenoalkanes and add $1\,cm^3$ ethanol to each. In turn add $2\,cm^3$ of aqueous silver nitrate, shake the test tube and time how long it takes to observe a precipitate of the silver halide — see core practical 4, page 29.

Enthalpy experiments

You are expected to be able to describe the techniques, draw suitable apparatus and discuss errors when measuring temperature changes for:

- rapid reactions in a polystyrene cup
- slow reactions in a polystyrene cup
- combustion using a spirit Lamp

For further details see pages 44–47.

Questions & Answers

This section contains multiple-choice and structured questions similar to those you can expect to find in AS paper 2 and in parts of questions in the A-level examinations. The questions given here are not balanced in terms of types of question or level of demand — they are not intended to typify real papers, only the sorts of questions that could be asked.

The answers given are those that examiners might expect from a top-grade student. They are not 'model answers' to be regurgitated without understanding. In answers that require more extended writing, it is usually the ideas that count rather than the form of words used. The principle is that correct and relevant chemistry scores.

Comments

Comments on the questions are indicated by the icon ⓔ. They offer tips on what you need to do to get full marks. Answers to questions might be followed by comments, preceded by the icon ⓔ, that explain the correct answer, point out common errors made by students who produce work of C-grade or lower, or contain additional material that could be useful to you.

The exam papers

Each AS paper lasts 1 hour 30 minutes and is worth 80 marks. A minimum of 20 marks will be awarded for the understanding of experimental skills and the interpretation of quantitative and qualitative experimental data. There will be 10–20 multiple-choice questions embedded in structured questions. In the A-level exams, questions on some or all the topics in this book will also be asked in papers 1 and 2 (both 1 hour 45 minutes with a maximum of 90 marks) and in paper 3 (2 hours 30 minutes with a maximum of 120 marks).

Assessment objectives

The AS and A-level exams have three assessment objectives (AOs). The percentages of each are very similar in the AS and the A-level papers, with a slightly heavier weighting of AO3 in the A-level.

AO1 is 'knowledge and understanding of scientific ideas, processes and procedures'. This makes up 36% of the exam. You should be able to:

- recognise, recall and show understanding of scientific knowledge
- select, organise and present information clearly and logically, using specialist vocabulary where appropriate

AO2 is 'application of knowledge and understanding of scientific ideas, processes and procedures'. This makes up 42% of the exam. You should be able to:

- analyse and evaluate scientific knowledge and processes
- apply scientific knowledge and processes to unfamiliar situations
- assess the validity, reliability and credibility of scientific information

AO3 is 'the analysis, interpretation and evaluation of scientific ideas and data'. This makes up 22% of the exam. You should be able to make scientific judgements, reach conclusions and evaluate and improve described practical procedures.

Command terms

The following command terms are used in the specification and in the AS and A-level papers. You must distinguish between them carefully.

- **Give, state or name** — no explanation is needed
- **Identify** — give the name or formula
- **Define/State what is meant by** — give a simple definition, without any explanation.
- **Describe** — state the characteristics of a particular material or process. No explanations are needed.
- **Explain** — use chemical theories or principles to say why a particular property of a substance or series of substances is as it is. It requires the making and justification of one or more points.
- **Deduce** — draw conclusions from information provided.
- **Calculate** — you are advised to show your working, so that consequential marks can be awarded even if you made a mistake in an earlier part of the calculation. The answer should include units and should be written to three significant figures unless the question asks for a specific or a suitable number of significant figures.

Revision

Start your revision in plenty of time. Make a list of the things that you need to do, emphasising the things that you find most difficult, and draw up a detailed revision plan. Work back from the examination date, ideally leaving an entire week free from fresh revision before that date. Be realistic in your revision plan and then add 25% to the timings because everything takes longer than you think.

When revising:

- make a note of difficulties and ask your teacher about them — if you do not make such notes, you will forget to ask
- make use of past papers, but remember that these may have been written to a different specification
- revise ideas, rather than forms of words — you are after *understanding*
- remember that scholarship requires time to be spent on the work

When you use the example questions in this book, make a determined effort to answer them before looking up the answers and comments. Remember that the answers given here are not intended as model answers to be learnt parrot-fashion. They are answers designed to illuminate chemical ideas and understanding.

The exam paper

- Read the question. Questions usually change from one examination to the next. A question that looks the same, at a cursory glance, as one that you have seen before usually has significant differences when read carefully. Needless to say, students do not receive credit for writing answers to their own questions.

- Be aware of the number of marks available for a question. That is a good indication of the number of things you need to say.
- Do not repeat the question in your answer. The danger is that you fill up the space available and think that you have answered the question, when in reality some or maybe all of the real points have been ignored.
- The name of a 'rule' is not an explanation for a chemical phenomenon. For example, in equilibrium, a common answer to a question on the effect of changing pressure on an equilibrium system is 'Because of Le Chatelier's principle…'. That is simply a name for a rule — it does not explain anything.

Multiple-choice questions

Answers to multiple-choice questions are machine-marked. Multiple-choice questions need to be read carefully; it is important not to jump to a conclusion about the answer too quickly. You need to be aware that one of the options might be a 'distracter'. An example of this might be in a question having a numerical answer of, say, $-600\,kJ\,mol^{-1}$; a likely distracter would be $+600\,kJ\,mol^{-1}$.

Some questions require you to think on paper — there is no demand that multiple-choice questions be carried out in your head. Space is provided on the question paper for rough working. It will not be marked, so do not write anything that matters in this space because no-one will see it.

For each of the questions there are four suggested answers, A, B, C and D. You select the best answer by putting a cross in the box beside the letter of your choice. If you change your mind you should put a line through the box and then indicate your alternative choice. Making more than one choice does not earn any marks. Note that this format is not used in the multiple-choice questions in this book.

Each test has at least ten multiple-choice questions, which are embedded in longer questions.

Online marking

It is important that you have some understanding of how examinations are marked, because to some extent it affects how you answer them. Your examination technique partly concerns chemistry and partly must be geared to how the examinations are dealt with physically. You have to pay attention to the layout of what you write. Because all your scripts are scanned and marked online, there are certain things you must do to ensure that all your work is seen and marked. These are covered below.

As the examiner reads your answer, decisions have to be made — is this answer worth the mark or not? Your job is to give the *clearest possible answer* to the question asked, in such a way that your chemical understanding is made obvious to the examiner. In particular, you must not expect the examiner to guess what is in your head; you can be judged only by what you write.

Because examination answers cannot be discussed, you must make your answers as clear as possible. This is one reason why you are expected, for example, to show working in calculations.

It is especially important that you *think before you write*. You will have a space on the question paper that the examiner has judged to be a reasonable amount for the answer. Because of differing handwriting sizes, false starts and crossings-out and because some students have a tendency to repeat the question in the answer space, the space is never exactly right for all students.

Edexcel exam scripts are marked online, so few examiners will handle a real, original script. The process is as follows:

- When the paper is set it is divided up into items, often, but not necessarily, a single part of a question. These items are also called clips.
- The items are set up so that they display on-screen, with check-boxes for the score and various buttons to allow the score to be submitted or for the item to be processed in some other way.
- After you have written your paper it is scanned; from that point all the handling of your paper is electronic. Your answers are tagged with an identity number.
- It is impossible for an examiner to identify a centre or a student from any of the information supplied.

Common pitfalls

The following list of potential pitfalls to avoid is particularly important:

- **Do not write in any colour other than black.** This is an exam board regulation. The scans are entirely black-and-white, so any colour used simply comes out black — unless you write in red, in which case it does not come out at all. The scanner cannot see red (or pink or orange) writing. So if, for example, you want to highlight different areas under a graph, or distinguish lines on a graph, you must use a different sort of shading rather than a different colour.
- **Do not use small writing.** Because the answer appears on a screen, the definition is slightly degraded. In particular, small numbers used for powers of 10 can be difficult to see. The original script is always available but it can take a long time to get hold of it.
- **Do not write in pencil.** Faint writing does not scan well.
- **Do not write outside the space provided without saying, within that space, where the remainder of the answer can be found.** Examiners only have access to a given item; they cannot see any other part of your script. So if you carry on your answer elsewhere but do not tell the examiner within the clip that it exists, it will not be seen. Although the examiner cannot mark the out-of-clip work, the paper will be referred to the principal examiner responsible for the paper.
- **Do not use asterisks or arrows as a means of directing examiners where to look for out-of-clip items.** Tell them in words. Students use asterisks for all sorts of things and examiners cannot be expected to guess what they mean.
- **Do not write across the centre-fold of the paper from the left-hand to the right-hand page.** A strip about 8 mm wide is lost when the papers are guillotined for scanning.
- **Do not repeat the question in your answer.** If you have a question such as 'Define the first ionisation energy of calcium', the answer is 'The energy change per mole for the formation of unipositive ions from isolated calcium atoms in the gas phase'; or, using the equation, 'The energy change per mole for $Ca(g) \rightarrow Ca^+(g) + e^-$'. Do not start by writing 'The first ionisation energy for calcium is defined as...' because by then you will have taken up most of the space available for the answer. Examiners know what the question is — they can see it on the paper.

∎Structured and multiple-choice questions

Question 1

This question is about alkanes and alkenes.

(a) Alkanes are members of a homologous series.

    (i) Explain the meaning of the term 'homologous series', using the alkanes as your example. (2 marks)

ⓔ Note that the question requires a reference to the homologous series of alkanes.

    (ii) Draw the displayed formulae of the three structural isomers of the alkane C_5H_{12}. (3 marks)

ⓔ Make sure that you write in all the hydrogen atoms. Using 'sticks' will not score full marks.

(b) Combustion of gasoline produces the power in petrol engines. One component of gasoline has the molecular formula C_8H_{18}.

    (i) The number of moles of oxygen needed in the equation that represents $\Delta_c H^\ominus$ of C_8H_{18} is: (1 mark)

      A 8

      B 12½

      C 17

      D 25

    (ii) Which substance is *not* produced in the exhaust of a petrol car? (1 mark)

      A H_2

      B H_2O

      C C

      D CO

(c) Methane reacts with chlorine in the presence of ultraviolet light.

    (i) Give the equation for the reaction that produces chloromethane. (1 mark)

    (ii) Write the mechanism of this reaction. (3 marks)

    (iii) Explain why some dichloromethane, CH_2Cl_2, is also produced when this reaction takes place. (2 marks)

ⓔ To answer this you must understand the mechanism. With which species, that will be present during the course of the reaction, could a chlorine radical or a methyl radical collide?

(d) Ethene is a member of the alkene homologous series. It reacts with bromine, but light is not needed.

    (i) Give the equation for the reaction and state what you would see. (2 marks)

    (ii) Give the mechanism for the reaction of an alkene with bromine. (3 marks)

(iii) 3-methylpent-2-ene is also an alkene and it has two geometric isomers. Giving your reasons, state whether the isomer shown in Figure 1 is the *E*-isomer or the *Z*-isomer.

(2 marks)

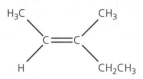

Figure 1

Total: 20 marks

Student answer

(a) (i) A series of compounds with the same general formula, in this case C_nH_{2n+2} ✓, and showing similar chemical properties ✓.

ℯ The inclusion of the particular general formula is necessary to make the answer refer to the alkanes, as required by the question. If asked about a specific type of compound, make sure that your answer is specific, not general.

(a) (ii)

$$H-\overset{\overset{\displaystyle H}{|}}{\underset{\underset{\displaystyle H}{|}}{C}}-\overset{\overset{\displaystyle H}{|}}{\underset{\underset{\displaystyle H}{|}}{C}}-\overset{\overset{\displaystyle H}{|}}{\underset{\underset{\displaystyle H}{|}}{C}}-\overset{\overset{\displaystyle H}{|}}{\underset{\underset{\displaystyle H}{|}}{C}}-\overset{\overset{\displaystyle H}{|}}{\underset{\underset{\displaystyle H}{|}}{C}}-H \quad ✓$$

$$H-\overset{\overset{\displaystyle H}{|}}{\underset{\underset{\displaystyle H}{|}}{C}}-\overset{\overset{\displaystyle H}{|}}{\underset{\underset{\displaystyle H}{|}}{C}}-\overset{\overset{\displaystyle H}{|}}{\underset{\underset{\displaystyle CH_3}{|}}{C}}-\overset{\overset{\displaystyle H}{|}}{\underset{\underset{\displaystyle H}{|}}{C}}-H \quad ✓$$

$$H-\overset{\overset{\displaystyle H}{|}}{\underset{\underset{\displaystyle H}{|}}{C}}-\overset{\overset{\displaystyle CH_3}{|}}{\underset{\underset{\displaystyle CH_3}{|}}{C}}-\overset{\overset{\displaystyle H}{|}}{\underset{\underset{\displaystyle H}{|}}{C}}-H \quad ✓$$

(b) (i) B (✓)

(ii) A (✓)

(c) (i) $CH_4 + Cl_2 \rightarrow CH_3Cl + HCl$ (✓)

(ii) Initiation: $Cl_2 + light \rightarrow 2Cl\bullet$ (✓)
Propagation:
$CH_4 + Cl\bullet \rightarrow \bullet CH_3 + HCl$
$\bullet CH_3 + Cl_2 \rightarrow CH_3Cl + Cl\bullet$ (✓)
Termination:
$2Cl\bullet \rightarrow Cl_2$
$Cl\bullet + \bullet CH_3 \rightarrow CH_3Cl$
$2\bullet CH_3 \rightarrow C_2H_6$ (✓)

ℯ A common error is to have an impossible propagation step such as $CH_4 + Cl\bullet \rightarrow CH_3Cl + H\bullet$. Not only are the bond energies unfavourable, but also the mechanism is the removal of an H radical by a Cl radical.

Questions & Answers

(c) (iii) A chlorine radical could collide with a CH_3Cl molecule produced earlier ✓. This will form some CH_2Cl_2 by the propagation steps:

$CH_3Cl + Cl\bullet \rightarrow \bullet CH_2Cl + HCl$

Then $\bullet CH_2Cl + Cl_2 \rightarrow CH_2Cl_2 + Cl\bullet$ ✓

ⓔ The reaction is not instantaneous, so some chlorine radicals will collide with CH_3Cl molecules produced earlier rather than with unreacted CH_4 molecules.

(d) (i) $H_2C=CH_2 + Br_2 \rightarrow BrCH_2CH_2Br$ ✓

The brown colour of bromine is lost. ✓

ⓔ Do *not* say that the bromine goes clear. Bromine is a clear brown colour. If bromine water is used, the product is CH_2BrCH_2OH, caused by the addition of HOBr, which is formed by the reaction $Br_2 + H_2O \rightarrow HOBr + HBr$.

(d) (ii)

ⓔ There is 1 mark for the first two arrows, 1 mark for the intermediate carbocation and 1 mark for the attack of the bromide ion on the carbocation. Note that curly arrows show the movement of electron pairs, so they must start and finish in the proper places. A curly arrow must start from a bond or from an atom or ion. It must go *to* an atom to form an ion or *towards* an atom to form a bond. Lone pairs should be shown but their omission may not be penalised.

(d) (iii) In the structure given the methyl group on the left-hand carbon atom takes precedence and the ethyl group on the right-hand carbon takes precedence ✓; these are across the double bond, so the isomer is the *E*-isomer ✓ (*entgegen*).

ⓔ Note that if this molecule is named using the *cis–trans* notation, it is the *cis*-isomer. The molecule is (*E*)-3-methylpent-2-ene.

Question 2

Consider the series of reactions in Figure 2, and then answer the questions that follow.

$$C_4H_8 \underset{\text{ethanolic KOH}}{\overset{\text{HBr}}{\rightleftharpoons}} C_4H_9Br \xrightarrow{\text{NaOH(aq)}} C_4H_9OH$$
A   B   C

Figure 2

(a) (i) Compound A is an alkene that has two geometric isomers. Draw their structural formulae.   (2 marks)

🄮 The fact that it has two geometric isomers eliminates two of the isomers of C_4H_8.

    (ii) State the *two* features of the molecule that makes geometric isomerism possible.   (2 marks)

    (iii) Stating what you would *see*, give a simple chemical test for the functional group present in compound A.   (2 marks)

🄮 Name the reagent, and the initial and final colours.

    (iv) Identify the *type* and the *mechanism* of the reaction converting compound A to compound B?   (2 marks)

    (v) Draw the mechanism for the reaction of compound A with hydrogen bromide. In your mechanism, the alkene (A) can be represented simply as:

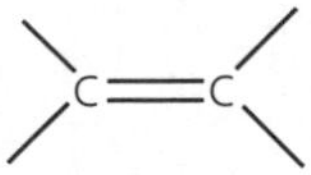

    Figure 3   (3 marks)

    (vi) What type of reaction is the conversion of compound B to compound A?   (1 mark)

(b) (i) The reaction of compound B to give compound C is a nucleophilic substitution. A nucleophile is a species that: (1 mark)

    A   donates an electron pair to form a covalent bond

    B   accepts an electron pair and forms a covalent bond

    C   must be negatively charged

    D   has a single unpaired electron

    (ii) Compound C is butan-2-ol. Its skeletal formula is: (1 mark)

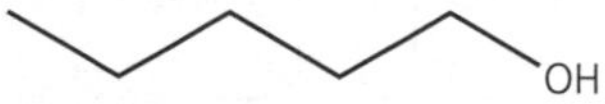

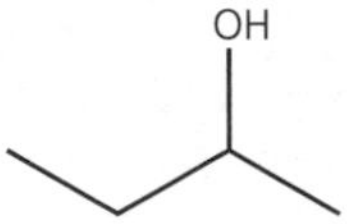

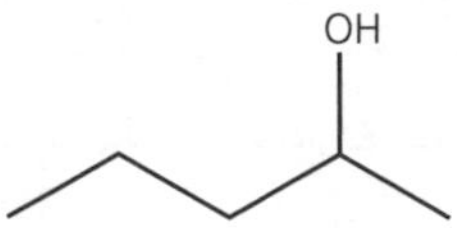

Questions & Answers

(b) (iii) Give a simple chemical test for the functional group in compound C. Describe what you would see as a result of this test. (2 marks)

(iv) Give the name and the displayed formula for the compound obtained by heating compound C with acidified potassium dichromate solution. (2 marks)

Total: 18 marks

Student answer

(a) (i)

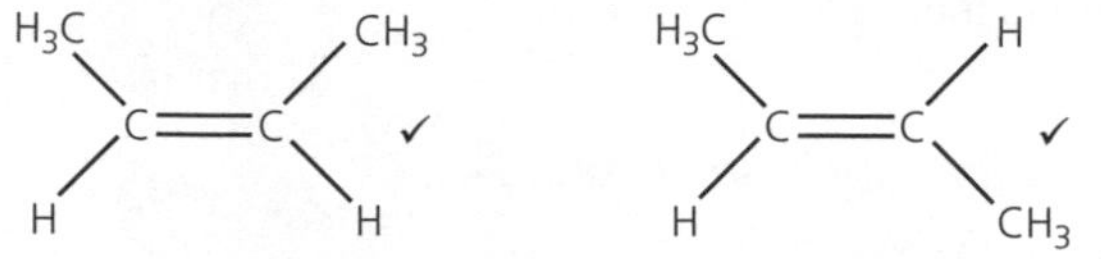

ⓔ Each correct structure scores 1 mark.

(a) (ii) Restricted (or no) rotation about the C=C bond ✓ and the two groups on a given carbon in the double bond are different ✓.

ⓔ The requirement that the groups on a given carbon must be different is often missed.

(a) (iii) Bromine water ✓ is decolorised from yellow (or orange) to colourless ✓.

ⓔ Note the difference between colourless (no colour) and clear, which means see-through and can be any colour. Confusion between the two is common. The fail-safe answer to all questions involving colour changes is to give the starting and finishing colours.

(a) (iv) Type of reaction: addition ✓; mechanism: electrophilic ✓

(v)

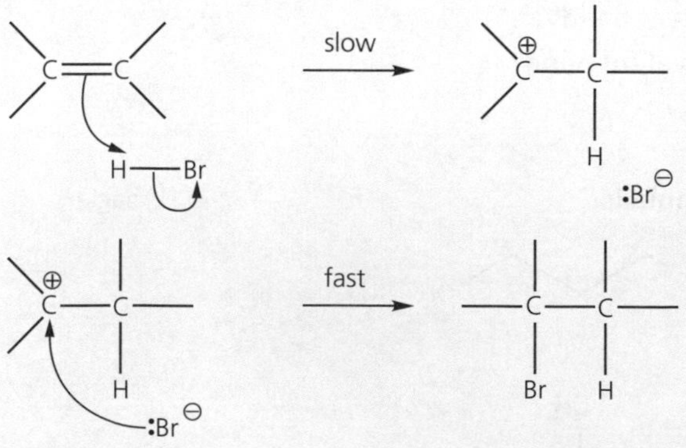

ⓔ The marks are for the two arrows initially ✓, for the structure of the carbocation intermediate ✓, and for the arrow in the second step ✓ to give the product.

The arrows show the movement of electron pairs and it is important to show their positions accurately. It is not essential to show the lone pair on the bromide ion, but it is better to include it because the lone pair makes the bromide a nucleophile. Similarly the labels 'slow' and 'fast' do not have to be added, but the statements are true so why not include them? Students seldom do — but it is knowledge of such details that indicates A-grade performance.

(a) (vi) Elimination ✓

(b) (i) A ✓

(ii) B ✓

e Formula A is butan-1-ol, and C and D are isomers of pentanol.

(b) (iii) Add phosphorus pentachloride ✓; steamy (acidic) fumes ✓

e 'Acid fumes' alone would not score because you cannot see that the fumes are acidic; nor are the fumes white, which is a common error.

(b) (iv) Name: butanone ✓

Formula:

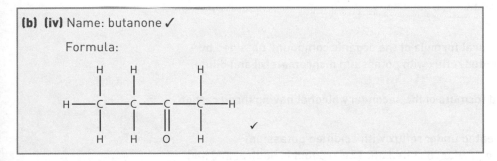

✓

e The name butan-2-one is acceptable, but the 2- is unnecessary as it is the only ketone of formula C_4H_8O.

In this question many of the parts depend on the previous part. So, if you think that compound A is but-1-ene, you might forget Markovnikoff's rule and think that compound B is a primary halogenoalkane and that compound C is a primary alcohol, which oxidises to an aldehyde or to a carboxylic acid. This would be taken into account in the mark scheme, so that marks could still be scored, even though an initial error had been made.

Question 3

Ethanal can be prepared by the oxidation of ethanol using potassium dichromate(vi) in dilute sulfuric acid, distilling off the ethanal as it is formed. The dichromate(vi) ions are reduced to chromium(iii) ions (Cr^{3+}).

$$Cr_2O_7^{2-} + 3CH_3CH_2OH + 8H^+ \rightarrow 2Cr^{3+} + 3CH_3CHO + 7H_2O$$

(a) The ability to distil out ethanal before it oxidises depends on its having a lower boiling temperature than ethanol. Explain in terms of the types of intermolecular forces present why an aldehyde always has a lower boiling temperature than the alcohol from which it can be made. (3 marks)

ⓔ List the intermolecular forces in *both* compounds and comment on their relative strengths.

(b) Ethanol can also be heated under reflux with potassium dichromate(vi) and dilute sulfuric acid.

(i) Explain what *heating under reflux* is and why it is used in preparative organic chemistry. (4 marks)

ⓔ You need to give a brief description of the apparatus and what happens during heating under reflux. In addition you must say why heating is needed and why under reflux.

(ii) Give the full structural formula of the organic compound obtained by heating ethanol under reflux with potassium dichromate(vi) and dilute sulfuric acid. (2 marks)

(iii) Give the displayed formula of the secondary alcohol having three carbon atoms. (1 mark)

(c) An alcohol C_3H_8O is heated under reflux with acidified potassium dichromate(vi), and the product distilled off. The product is neutral and had the formula C_3H_6O and it had no effect on Fehling's solution. Which of the following statements is true? (1 mark)

A The alcohol is a primary alcohol and it has been oxidised to an aldehyde.

B The alcohol is a secondary alcohol and it has been oxidised to a ketone.

C The alcohol is a primary alcohol and it has been oxidised to a carboxylic acid.

D The alcohol is a tertiary alcohol.

Total: 11 marks

> ### Student answer
>
> (a) The alcohol has intermolecular hydrogen bonds ✓, whereas the aldehyde has only London (dispersion) and dipole–dipole forces ✓. Hydrogen bonding is the strongest so more energy is needed to overcome it ✓.

ⓔ Note that when you answer questions concerning two compounds, you need to ensure that you mention both of them.

(b) (i) The reaction flask is fitted with a vertical condenser ✓, which returns the condensed liquid continuously to the flask ✓. The reaction is slow and so needs heating ✓ and use of a reflux condenser enables prolonged heating without loss of volatile reactant ✓.

(ii)

$$CH_3 - C \begin{matrix} O \\ \\ OH \end{matrix}$$

ⓔ 1 mark for giving a carboxylic acid and 1 mark for giving ethanoic acid.

(iii)

$$H - \overset{\displaystyle H}{\underset{\displaystyle H}{C}} - \overset{\displaystyle H}{\underset{\displaystyle H}{C}} - \overset{\displaystyle H}{\underset{\displaystyle O}{C}} - \overset{\displaystyle H}{\underset{\displaystyle H}{C}} - H$$ ✓

ⓔ The bond in the O–H group is shown because a full structural formula is required. Showing the group as –OH would probably be accepted, but it is best not to risk this.

(c) B

ⓔ An aldehyde would react with Fehling's solution. A carboxylic acid is not neutral and tertiary alcohols are not oxidised.

Question 4

Spectroscopy is a powerful tool in the determination of molecular structure. Mass spectroscopy and infrared spectroscopy are widely used with organic molecules.

(a) An organic compound X is believed to be either butanone or butan-2-ol. Its mass spectrum is shown in Figure 4.

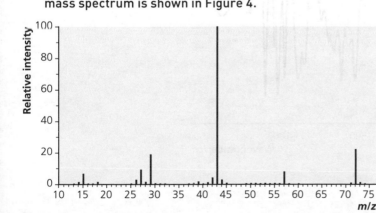

Figure 4

    (i) Explain why, on the basis of the peak at m/z 72, the compound is butanone rather than butan-2-ol.   (2 marks)

ℯ You must make a comment about both substances.

    (ii) Give the formula of the species that is responsible for the peak at m/z 29.   (2 marks)

    (iii) Write the equation for the reaction that produces the fragment at m/z 43 from the molecular ion.   (2 marks)

(b) Infrared spectroscopy can readily distinguish alcohols such as butan-2-ol from ketones such as butanone.

    (i) State what property a bond must possess if it is to absorb in the infrared region of the spectrum.   (1 mark)

    (ii) Which of the following statements about the oxidation of an alcohol, such as $C_4H_{10}O$, is true?   (1 mark)

        A The alcohol has been oxidised to a ketone if the broad O–H absorption at about $3300\,cm^{-1}$ in the alcohol spectrum is replaced by a narrow C=O absorption at about $1720\,cm^{-1}$ in the product spectrum.

        B The alcohol has been oxidised to an aldehyde if the narrow O–H absorption at $3300\,cm^{-1}$ in the alcohol spectrum is replaced by a broad C=O absorption at about $1720\,cm^{-1}$ in the product spectrum.

        C You cannot tell whether the alcohol has been oxidised to a ketone or an aldehyde because these compounds do not absorb in the IR part of the spectrum.

        D You cannot tell whether the alcohol has been oxidised to a ketone or an aldehyde because both ketones and aldehydes give an IR absorption at around $1720\,cm^{-1}$.

(c) Use the data booklet and the spectrum shown in Figure 5 to answer the following questions.

    (i) Identify the bond and mode of vibration responsible for the peaks at $2900\,cm^{-1}$ and at $3350\,cm^{-1}$.   (2 marks)

    (ii) Explain why the peak at $3350\,cm^{-1}$ is broad.   (1 mark)

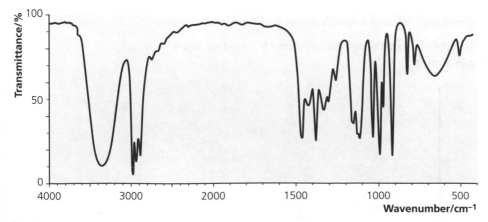

Figure 5

Total: 11 marks

> **Student answer**
>
> **(a) (i)** The peak at $m/z = 72$ is the molecular ion peak and corresponds to the relative molecular mass of butanone ✓. Butan-2-ol would have a molecular ion peak at $m/z = 74$ ✓.
>
> **(ii)** $m/z = 29$ is $CH_3CH_2^+$

ⓔ 1 mark for CH_3CH_2 and 1 mark for the + charge. Weaker students often forget to include the charge, costing them a mark.

> **(iii)** $CH_3COC_2H_5^+ \rightarrow C_2H_5\bullet$ ✓ $+ CH_3CO^+$ ✓
>
> **(b) (i)** The bond must change its polarity when it stretches or bends ✓.
>
> **(ii)** D ✓

ⓔ The frequencies of the C=O stretch in both aldehydes and ketones are very similar.

> **(c) (i)** The peak at $2900\,cm^{-3}$ is caused by C–H stretching ✓ and that at $3350\,cm^{-1}$ by O–H stretching ✓.
>
> **(ii)** The peak is broad due to hydrogen bonding ✓.

Question 5

(a) (i) Define the term 'standard enthalpy of combustion'. (3 marks)

ⓔ Do not forget to state what is meant by 'standard conditions'.

(ii) Which one of the following equations is associated with the definition of the standard enthalpy of formation of carbon monoxide? (1 mark)

A $C(g) + \frac{1}{2}O_2(g) \rightarrow CO(g)$

B $C(graphite) + \frac{1}{2}O_2(g) \rightarrow CO(g)$

C $C(graphite) + O(g) \rightarrow CO(g)$

D $C(g) + O(g) \rightarrow CO(g)$

(iii) If the average C–H bond enthalpy is $+x$, which one of the following represents a process with enthalpy change $+4x$? (1 mark)

A $C(g) + 4H(g) \rightarrow CH_4(g)$

B $CH_4(g) \rightarrow C(g) + 4H(g)$

C $CH_4(g) \rightarrow C(g) + 2H_2(g)$

D $C(s) + 2H_2(g) \rightarrow CH_4(g)$

(iv) State Hess's law. (2 marks)

(v) Draw a Hess's law diagram and use it to calculate the enthalpy change of formation of ethane, given: (3 marks)

	C(s)	H_2(g)	C_2H_6(g)
$\Delta_c H$/kJ mol^{-1}	–394	–286	–1560

Questions & Answers

(b) The enthalpy change for the reaction $CuSO_4.5H_2O(s) \rightarrow CuSO_4(s) + 5H_2O(l)$ can be found using Hess's law and measuring the $\Delta_{soln}H^{\ominus}$ of both anhydrous and hydrated copper sulfate. You have been asked to plan an experiment that would enable you to measure $\Delta_{soln}H^{\ominus}$ of anhydrous copper sulfate.

Describe exactly how you would carry out this experiment, including describing any precautions that you would make to ensure an accurate result. You do *not* need to show how you would calculate the value of $\Delta_{soln}H^{\ominus}$ from your data.

(7 marks)

(c) The equation for the combustion of ethanol in air is:

$C_2H_5OH(l) + 3O_2(g) \rightarrow 2CO_2(g) + 3H_2O(l)$

Calculate the enthalpy change for this reaction using the average bond enthalpy values given in Table 1.

(3 marks)

Bond	Average bond enthalpy/kJ mol^{-1}	Bond	Average bond enthalpy/kJ mol^{-1}
C–H	+412	C–C	+348
C–O	+360	O–H	+463
O=O	+496	C=O	+743

Table 1

Total: 20 marks

ⓔ Make sure that you give a sign and units with your answer.

> **Student answer**
>
> **(a) (i)** The heat energy change per mole ✓ for the complete combustion of a substance in excess oxygen ✓ at 100 kPa pressure (allow 1 atm) and stated temperature ✓.

ⓔ In practice, quoting the temperature as 298 K would also gain credit, although this particular temperature is not part of the definition of the standard state. *Heat* energy change, not energy change alone, is important. The energy change for a reaction is given the symbol ΔU and is measured at constant volume, not constant pressure.

> **(ii)** B

ⓔ Bond enthalpies refer to breaking the bond and forming atoms.

> **(iii)** B

ⓔ The answer must relate to the most stable form of carbon, so it is B not A.

> **(iv)** The heat energy/enthalpy change in a chemical reaction is independent of the route used to go from the reagents to the products ✓ provided that the initial and final states are the same ✓.

e The term 'energy change' alone should not be used — this refers to a reaction at constant volume, not constant pressure.

(v)

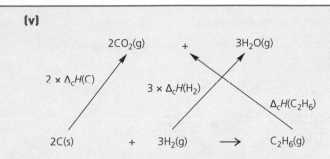

$$\Delta_f H(C_2H_6) = 2 \times \Delta_c H(C) + 3 \times \Delta_c H(H_2) - \Delta_c H(C_2H_6)$$

$$= (-788) + (-858) - (-1560) = -86\,kJ\,mol^{-1}$$

Diagram ✓ value ✓ sign and units ✓

(b) Weigh an empty weighing bottle and then containing some anhydrous copper sulfate ✓.

Using a measuring cylinder (allow pipette) measure out a known volume (allow a stated number) of water into a polystyrene cup ✓.

Measure the temperature of the water every minute for 4 minutes ✓.

On the fifth minute, add the copper sulfate to the water and stir ✓.

Measure the temperature (and continue to stir) every minute until all the solid has dissolved ✓.

Plot a graph of temperature against time and extrapolate both lines to 5 minutes and read off the rise in temperature ✓.

Make sure you read the balance to 0.01 g and the temperature to at least 0.5°C ✓.

(c) There is no standard notation for the average bond enthalpy. Using $D(X–X)$ for the X–X bond:

ΔH = (sum of bond energies of reagents) – (sum of bond energies of the products) ✓

$= [5D(C–H) + D(C–O) + D(C–C) + D(O–H) + 3D(O=O)] - [4D(C=O) + 6D(O–H)]$ ✓

$= [(5 \times 412) + 360 + 348 + 463 + (3 \times 496)] - [(4 \times 743) + (6 \times 463)] = -1031\,kJ\,mol^{-1}$ ✓

e Note that brackets are used to separate the enthalpy values for each type of bond. The absence of such working is characteristic of C-grade answers and is a high-risk strategy. If the answer only is shown and it is wrong, you cannot get marks for intermediate steps.

Remember that bond making is exothermic and bond breaking endothermic.

Be careful to choose the correct values — it is common to find, for example, that the C–O bond strength is used for the C=O bond.

Question 6

The reaction between sulfur dioxide and oxygen in a closed system is in dynamic equilibrium:

$$2SO_2(g) + O_2(g) \rightleftharpoons 2SO_3(g) \qquad \Delta_rH = -196\,kJ\,mol^{-1}$$

(a) (i) Explain what is meant by *dynamic equilibrium*. (2 marks)

ⓔ Remember to define both *dynamic* and *equilibrium*.

(ii) The correct expression for K_c for this reaction is: (1 mark)

A $\dfrac{[SO_3]}{[SO_2]\,[O_2]^{\frac{1}{2}}}$ B $\dfrac{[SO_3]^2}{[SO_2]^2\,[O_2]}$ C $\dfrac{[SO_2]\,[O_2]^{\frac{1}{2}}}{[SO_3]}$ D $\dfrac{[SO_2]^2\,[O_2]}{[SO_3]^2}$

(b) State the effect on the position of equilibrium of this reaction of:

(i) increasing the temperature (1 mark)

(ii) increasing the pressure (1 mark)

ⓔ There is no need for any explanation.

(c) This reaction is the first step in the industrial production of sulfuric acid. A temperature of 450°C, a pressure of 2 atm and a catalyst are used. Justify the use of these conditions:

(i) a temperature of 450°C (2 marks)

(ii) a pressure of 2 atm (2 marks)

(iii) a catalyst (1 mark)

ⓔ For each, a comment on the *rate* and another on the *yield* is needed.

(d) Name the catalyst used industrially. (1 mark)

(e) Explain why an industrial manufacturing process such as this cannot be in equilibrium. (2 marks)

(Total: 13 marks)

Student answer

(a) (i) The forward and backward reactions are occurring at the same rate ✓, so there is no net change in composition ✓.

ⓔ The first mark is for *dynamic* and the second for *equilibrium*.

(ii) B

ⓔ A is for the equation $SO_2 + \frac{1}{2}O_2 \rightleftharpoons SO_3$. C and D are for the reverse reactions.

(b) (i) Moves to the left *or* concentration of products decreases ✓

(ii) Moves to the right *or* concentration of products increases ✓

ⓔ The equilibrium mixture does not move in a physical sense. The terms are used as shorthand — 'moves to the right' means that the concentrations of the substances on the right-hand side of the equation (the products, conventionally) increase. The effect of an increase in pressure on the rate of a heterogeneously catalysed reaction is negligible. The reactants have a strong affinity for the surface of the catalyst. At all pressures (other than extremely low), the catalyst surface, which is where the reaction occurs, is saturated with the reactants and so increasing the pressure makes no difference. This point is commonly misunderstood.

> **(c) (i)** The temperature needs to be high to give an acceptable rate but low to achieve an acceptable yield ✓; the temperature used is therefore a compromise *or* the temperature is the optimum for the catalyst used ✓.

ⓔ When the equilibrium temperature is increased, the equilibrium composition changes so that heat energy is absorbed, but this does not bring the temperature down again. The changes are to the *equilibrium* conditions — the alteration in composition has no effect on the conditions, which are externally imposed.

> **(ii)** High pressure increases the yield but is not necessary since an acceptable yield is obtainable at low pressures; *or* you need a high enough pressure to drive the gases through the catalyst bed(s) ✓; an increase in pressure would increase yield but this would not offset the increase in costs ✓.

ⓔ The yield is high because, in practice, four catalyst beds are used at different temperatures and these are efficient enough at low pressure. High-pressure sulfur trioxide is extremely corrosive. The use of high pressure would also cause the sulfur dioxide to liquefy in those parts of the plant that are at low temperature. The biggest cost in using high pressure is the cost of fuel for compressing the gases.

> **(iii)** Increases reaction rate but does not affect the yield ✓.

ⓔ The implication from a cost–yield viewpoint is that the rate becomes acceptable at a lower temperature than would otherwise be the case.

> **(d)** Vanadium(v) oxide *or* vanadium pentoxide ✓

ⓔ Platinum can be used in the laboratory, but it is not used industrially, so it is not an acceptable answer.

> **(e)** Products are being removed to be sold ✓ so the system is not closed and the reverse reaction cannot occur quickly enough to achieve equilibrium ✓.

ⓔ A closed system is one in which no exchange of matter is possible with the surroundings. Reactions in test tubes are in open systems since gases can be evolved. Equilibrium ideas can be used to approximately predict optimum conditions in industrial processes, but these processes cannot be in closed systems.

Question 7

The rate of a chemical reaction increases as the temperature is increased.

(a) Draw a diagram of the Maxwell–Boltzmann distribution of molecular energies at a temperature T_1 and at a higher temperature T_2. (3 marks)

ⓔ Don't forget to label the axes and the two curves.

(b) (i) Mark *on your diagram* in a suitable position the activation energy and use this to explain why the rate of a chemical reaction increases with increasing temperature. (4 marks)

ⓔ Make sure that you refer to the diagram. The increase in collision frequency can be ignored as it is far less important than the change in collision energy.

(ii) A catalyst increases the rate of a chemical reaction by: (1 mark)

A lowering the activation energy for the reaction

B giving the reacting molecules more energy so more of them collide successfully per unit time

C providing an alternative pathway with an activation energy that is lower than that of the uncatalysed reaction

D providing a surface on which the reactants can combine

(Total: 8 marks)

Student answer

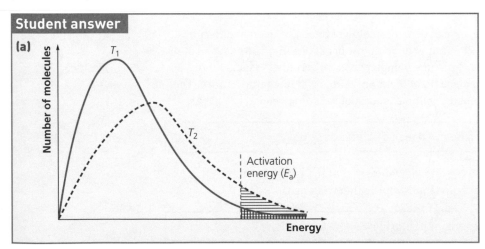

ⓔ Correctly labelled axes ✓; curve for T_1 skewed and correct shape within reason (starts at the origin and does not intercept the x-axis at high energy) ✓; curve for T_2 with peak to the right and lower ✓.

The Maxwell–Boltzmann curve has a well-defined shape and you are expected to be able to reproduce it accurately. Common errors include drawing a curve that is symmetrical or a curve that either starts part way up the y-axis or intersects the x-axis. The Maxwell–Boltzmann curve is not like a normal distribution (bell) curve.

(b) (i) E_a shown on the diagram well to the right of the peak ✓; area under the curve for energies above E_a is higher for T_2 than for T_1 ✓, so there is a higher proportion of successful (✓) collisions, since collision energy $> E_a$ for these molecules ✓.

e The idea of collision energy is important, as is the ability to see kinetics in terms of particle interactions. If you do not refer to the diagram, you can earn a maximum of only 3 marks.

The higher proportion of successful collisions is an important, but subtle, point. It is not true to say, as many students do, that the number of collisions increases. Overall, the number of collisions leading to reaction for the same amount of product is the same. It is the time during which these collisions occur that matters. You could also say that the number of successful collisions per unit time increases.

Remember that your script is scanned before marking, so different colours cannot be distinguished and red is not picked up. When you shade the areas under the curves, use different types of shading or hatching, not colours.

(ii) C ✓

e The essential feature of a catalyst is that it changes the mechanism of a reaction to one that has lower activation energy than the uncatalysed reaction. The activation energy for a particular reaction mechanism is intrinsic to that mechanism and cannot be changed. Option D is often true, but does not apply to all catalysts.

Knowledge check answers

1 A secondary alcohol and an alkene

2 2,2-dimethylbutane

3 3,4-dimethylhex-2-ene

4

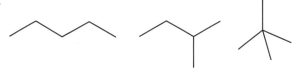

5 There are two hydrogen atoms on the left-hand carbon atom.

6

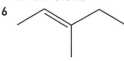

7

8

9 a 2-bromo-2-methylpropane, as it is tertiary and 1-bromo-2-methylpropane is primary.

  b 2-bromopropane, as the C–Br bond is weaker than the C–Cl bond.

10 Pass the product into bromine water. It changes from orange-brown to colourless.

11 2-methylpropan-1-ol is a primary alcohol, so the solution would go from orange to green. 2-methylpropan-2-ol is a tertiary alcohol and so will not be oxidised. The solution will stay orange.

12 Ethanal has no $\delta+$ H atom, so cannot form hydrogen bonds. It can only form London (dispersion) and dipole–dipole forces, which are weaker than the hydrogen bonds between ethanol molecules.

13 (1) The use of an expensive reagent such as sodium dichromate(VI) would be too expensive on an industrial scale, or, disposal of toxic chromium compounds is expensive. (2) An impure aqueous solution of ethanal would require expensive separation, whereas the vapour-phase oxidation gives the pure aldehyde.

14 The $CH_3CH_2^+$ ion and the $(CH_3)_2CH^+$ ion (caused by loss of C_2H_5)

15 The CH_2OH^+ ion

16 The peak at between 3500 and $3300\,cm^{-1}$ (there is no N–H bond in trimethylamine).

17 Error = $2 \times 0.5 = 1°C$
$\Delta T = 26.0 - 18.5 = 7.5°C$
% error = $1 \times 100/7.5 = 13\%$

18 Heat change = $mc\Delta T = 100\,g \times 4.2\,J\,K^{-1}g^{-1} \times 15\,K = 6300\,J$
Amount of ethanol = mass/molar mass = $0.23\,g/46\,g\,mol^{-1} = 0.005\,mol$
$\Delta H = -6300\,J/0.005\,mol = -1\,260\,000\,J\,mol^{-1} = -1260\,kJ\,mol^{-1}$

19 Equation 1: $CuSO_4.5H_2O(s) \rightarrow CuSO_4(aq)$
$\Delta H = +6\,kJ\,mol^{-1}$
Equation 2: $CuSO_4(s) \rightarrow CuSO_4(aq)$ $\Delta H = -73\,kJ\,mol^{-1}$
Reversing equation 2 and adding to equation 1 gives:
$CuSO_4.5H_2O(s) \rightarrow CuSO_4(s)\,(+5H_2O(l))$
$\Delta H = +6 - (-73) = +79\,kJ\,mol^{-1}$

20

Bonds broken	Value/ $kJ\,mol^{-1}$	Bonds made	Value/ $kJ\,mol^{-1}$
C=C	612	C–H	413
H–Cl	432	C–Cl	346
		C–C	347
Total	1044	Total	1106

$\Delta H_{reaction} = 1044 - 1106 = -62\,kJ\,mol^{-1}$
Do not forget that a new C–C has to be made as the calculation assumed that the C=C was completely broken.

21 The solution goes green (a mixture of blue and yellow) and then yellow with excess HCl. This is because the equilibrium is driven to the right by an increase of Cl^- ions from the HCl.

22 The gaseous mixture initially goes browner as the gas is compressed into a smaller volume and then slowly gets paler as the equilibrium shifts to the left (the side with fewer moles of gas).

23 $K_c = \dfrac{[CO]\,[H_2]^3}{[CH_4]\,[H_2O]}$

24 $K_c = [Ag^+]^2[SO_4^{2-}]$

25 The ether is highly flammable and because it is very volatile the hazard is considerable, so warm the boiling tube with a beaker of hot water rather than a Bunsen flame.

26 First hazard: concentrated sulfuric acid is corrosive. Risk reduced by wearing gloves.
Second hazard: gaseous bromine is toxic and harmful when breathed. Risk reduced by carrying out the experiment in a fume cupboard.

Index

Index

The periodic table

Group

Key:

| Relative atomic mass |
| **Atomic symbol** |
| name |
| Atomic (proton) number |

Period	1	2												3	4	5	6	7	0
1	1.0 **H** hydrogen 1																		4.0 **He** helium 2
2	6.9 **Li** lithium 3	9.0 **Be** beryllium 4												10.8 **B** boron 5	12.0 **C** carbon 6	14.0 **N** nitrogen 7	16.0 **O** oxygen 8	19.0 **F** fluorine 9	20.2 **Ne** neon 10
3	23.0 **Na** sodium 11	24.3 **Mg** magnesium 12												27.0 **Al** aluminium 13	28.1 **Si** silicon 14	31.0 **P** phosphorus 15	32.1 **S** sulfur 16	35.5 **Cl** chlorine 17	39.9 **Ar** argon 18
4	39.1 **K** potassium 19	40.1 **Ca** calcium 20	45.0 **Sc** scandium 21	47.9 **Ti** titanium 22	50.9 **V** vanadium 23	52.0 **Cr** chromium 24	54.9 **Mn** manganese 25	55.8 **Fe** iron 26	58.9 **Co** cobalt 27	58.7 **Ni** nickel 28	63.5 **Cu** copper 29	65.4 **Zn** zinc 30	69.7 **Ga** gallium 31	72.6 **Ge** germanium 32	74.9 **As** arsenic 33	79.0 **Se** selenium 34	79.9 **Br** bromine 35	83.8 **Kr** krypton 36
5	85.5 **Rb** rubidium 37	87.6 **Sr** strontium 38	88.9 **Y** yttrium 39	91.2 **Zr** zirconium 40	92.9 **Nb** niobium 41	95.9 **Mo** molybdenum 42	[98] **Tc** technetium 43	101.1 **Ru** ruthenium 44	102.9 **Rh** rhodium 45	106.4 **Pd** palladium 46	107.9 **Ag** silver 47	112.4 **Cd** cadmium 48	114.8 **In** indium 49	118.7 **Sn** tin 50	121.8 **Sb** antimony 51	127.6 **Te** tellurium 52	126.9 **I** iodine 53	131.3 **Xe** xenon 54
6	132.9 **Cs** caesium 55	137.3 **Ba** barium 56	138.9 **La** lanthanum 57	178.5 **Hf** hafnium 72	180.9 **Ta** tantalum 73	183.8 **W** tungsten 74	186.2 **Re** rhenium 75	190.2 **Os** osmium 76	192.2 **Ir** iridium 77	195.1 **Pt** platinum 78	197.0 **Au** gold 79	200.6 **Hg** mercury 80	204.4 **Tl** thallium 81	207.2 **Pb** lead 82	209.0 **Bi** bismuth 83	[209] **Po** polonium 84	[210] **At** astatine 85	[222] **Rn** radon 86
7	[223] **Fr** francium 87	[226] **Ra** radium 88	[227] **Ac** actinium 89	[261] **Rf** rutherfordium 104	[262] **Db** dubnium 105	[266] **Sg** seaborgium 106	[264] **Bh** bohrium 107	[277] **Hs** hassium 108	[268] **Mt** meitnerium 109	[271] **Ds** darmstadtium 110	[272] **Rg** roentgenium 111							

Elements with atomic numbers 112–116 have been reported but not fully authenticated

140.1 **Ce** cerium 58	140.9 **Pr** praseodymium 59	144.2 **Nd** neodymium 60	144.9 **Pm** promethium 61	150.4 **Sm** samarium 62	152.0 **Eu** europium 63	157.2 **Gd** gadolinium 64	158.9 **Tb** terbium 65	162.5 **Dy** dysprosium 66	164.9 **Ho** holmium 67	167.3 **Er** erbium 68	168.9 **Tm** thulium 69	173.0 **Yb** ytterbium 70	175.0 **Lu** lutetium 71
232 **Th** thorium 90	[231] **Pa** protactinium 91	238.1 **U** uranium 92	[237] **Np** neptunium 93	[242] **Pu** plutonium 94	[243] **Am** americium 95	[247] **Cm** curium 96	[245] **Bk** berkelium 97	[251] **Cf** californium 98	[254] **Es** einsteinium 99	[253] **Fm** fermium 100	[256] **Md** mendelevium 101	[254] **No** nobelium 102	[257] **Lr** lawrencium 103